Healthy Buildings Asia 2025

Hyderabad, India
18 - 21 June 2025

ISBN: 979-8-3313-3256-3

Printed from e-media with permission by:

Curran Associates, Inc.
57 Morehouse Lane
Red Hook, NY 12571

Some format issues inherent in the e-media version may also appear in this print version.

Printed with permission by Curran Associates, Inc. (2026)

For permission requests, please contact International Society of Indoor Air Quality and Climate (ISIAQ) at the address below.

International Society of Indoor Air Quality and Climate (ISIAQ)
c/o Infinity Conference Group, Inc.
1035 Sterling Road, Suite 202
Herndon, VA 20170 USA

Phone: (703)-925-9455 x118

info@isiaq.org

Additional copies of this publication are available from:

Curran Associates, Inc.
57 Morehouse Lane
Red Hook, NY 12571 USA
Phone: 845-758-0400
Fax: 845-758-2633
Email: curran@proceedings.com
Web: www.proceedings.com

Healthy Buildings Asia 2025

Healthy Buildings Asia 2025
Hyderabad, India, August 18-21, 2025

ISIAQ INTERNATIONAL SOCIETY OF INDOOR AIR QUALITY AND CLIMATE

HEALTHY BUILDINGS - 2025
ASIA · AFRICA · AMERICA · AUSTRALIA
THE FLAGSHIP PROGRAM OF THE INTERNATIONAL SOCIETY OF INDOOR AIR QUALITY AND CLIMATE

International Society of Indoor Air Quality and Climate (ISIAQ)
Healthy Buildings is a flagship conference series sponsored by ISIAQ

SPECIAL THANKS TO OUR PARTNERS

PARTNERS FOR SHARING FOR A BETTER FUTURE INITIATIVE

Drexel University Dornsife School of Public Health

LABVANCED

techknowgreen solutions limited

The Corsi-Rosenthal Foundation

PODCAST PARTNER

Air Quality Matters Podcast

SOCIETY PARTNER

ASHRAE Deccan Chapter

ISHRAE

FOCUS

CEAI Consulting Engineers Association of India

ASHRAE Mumbai Chapter

SIE

India Insulation Forum

KNOWLEDGE PARTNERS

aeon NOIDA

AHRI AIR-CONDITIONING, HEATING, & REFRIGERATION INSTITUTE USA

CARBSE CENTER FOR ADVANCED RESEARCH IN BUILDING SCIENCE AND ENERGY AHMEDABAD

Mahindra University | CENTRE FOR SUSTAINABILITY HYDERABAD

CVB - India

eds environmental design solutions NEW DELHI

GLOBAL CENTRE FOR CLEAN AIR RESEARCH UNIVERSITY OF SURREY ENGLAND

HOBEL HUMAN ORIENTED BUILT ENVIRONMENT LAB SWITZERLAND

Human Building-Interaction Lab IIT Kharagpur

chha RESPECT INNOVATION SKILLS EMPATHY HYDERABAD

MUMBAI

HYDERABAD

IGBC HYDERABAD

IIT INDIAN INSTITUTE OF TECHNOLOGY VARANASI

DELHI

iAIR INSTITUTE USA

IEQ LAB SYDNEY

ISAC Centre for Built Environment Policy MYSORE

THE FUTURE OF URBAN X PUNE

ktu 1922 faculty of civil engineering and architecture LITHUANIA

UNIVERSITY OF LEEDS ENGLAND

NIRMA UNIVERSITY NAAC ACCREDITED 'A+' GRADE AHMEDABAD

RICH Research and Innovation Circle of Hyderabad

U.S. GREEN BUILDING COUNCIL USGBC

SPECIAL THANKS TO OUR SPONSORS

Safe-air UK

AAF INTERNATIONAL
a member of DAIKIN group

testo

ASTBERG VENTILATION

Sustainable Stewardship

ALMED EQUIPMENT PVT LTD

airinspace

IICRC
Institute of Inspection Cleaning and Restoration Certification

camfil

Vinyas ELECTRONICS
Partner of TSI

Prof. Ranga Reddy Burri
Honorary President

Dr Reddy is a Physician, Public Health Policy Advisor, Academician, and Social Entrepreneur. He is an Honorary Professor at the University of Hyderabad and an Adjunct Professor at the Public Health Foundation of India.
His work focuses on public health awareness, education, and training. He is committed to improving Infection Prevention practices and patient safety in India and other low-resource settings. He was trained in the fields of medicine, management, public health and health at several prestigious universities including Minsk State Medical Institute, Pondichery University, IESE Business School, Washington University and Calgary University.

Vinod Kumar Sekar
President

ISIAQ member since 2016. Served as ISIAQ conference selection committee member. Pursuing a Doctorate in microbiology. Director of Airindoor Assessment Pvt Ltd has 15 years of experience in Indoor air and environmental quality audits and training. Involved in research and product developments to control indoor air and environmental microbial load. Member of R&D in LM laborites in India, Nepal BioScience Research Laboratory in Nepal and Bedair Ltd, UK.

Patrons

Mukesh Khare
Professor Emeritus,
Civil Engineering
Department,
IIT Delhi

R Chandrashekhar
Former Chief Architect,
Ministry of Health &
Family Welfare,
Govt. of India

G Rameshwar Rao
Director
Engineering Staff College
of India
Hyderabad

Scientific Committee

Richard de Dear, Chair of the Scientific Committee

Professor and Director, Indoor Environmental Quality Laboratory,
School of Architecture, Design and Planning,
The University of Sydney

Name	Designation & Institute	Country
Chirag Deb	Lecturer in Building Science, School of Architecture, Design and Planning, The University of Sydney	Australia
Ardeshir Mahdavi	Professor, Institute of Building Physics, Services, and Construction, Faculty of Civil Engineering Sciences, TU Graz	Austria
Jelle Laverge	Assoc. Prof., Department of Architecture and Urban planning, Ghent University	Belgium
Bin Cao	Associate Professor, Department of Building Science, School of Architecture, Tsinghua University	China
Bin Yang	Prof. Dean School of Energy and Safety Engineering, Tianjin Chengjian University	China
Da Yan	Professor, Editor-in-chief of International Journal of Building Simulation, School of Architecture, Tsinghua University	China
Jianbang Xiang	Assoc. Prof., School of Public Health (Shenzhen), Sun Yat-sen University	China
Nan Zhang	Assistant Researcher, Department of Building Science, School of Architecture, Tsinghua University	China
Ying Xu	Guo Qiang Distinguished Associate Professor, Department of Building Science, Tsinghua University Beijing, | Adjunct Associate Professor, Department of Civil, Architectural and Environmental Engineering, The University of Texas at Austin, Austin, U.S.A.	China
Yinping Zhang	Prof. & Director, Beijing Key Lab of Indoor Air Quality, Department of Building Science, Tsinghua University, President of Academy of ISIAQ Fellows	China
Alireza Afshari	Professor, Department of the Built Environment (BUILD), Aalborg University, Copenhagen	Denmark
Pawel Wargocki	Professor, International Centre for indoor Environment and Energy, Department of Environmental and Resource Engineering, Technical University of Denmark	Denmark
Hannu Salmela	VTT Technical Research Centre of Finland	Finland
Kati Huttunen	Senior researcher, Indoor Environments Team, Environmental Health Unit, Department of Health Security, Finnish Institute for Health and Welfare (THL)	Finland
Corinne Mandin	Head of the Laboratory of Epidemiology, Institute for Radiation Protection and Nuclear Safety	France

Chun CHEN	Associate Professor, Department of Mechanical and Automation Engineering, Institute of Environment, Energy, and Sustainability (IEES), The Chinese University of Hong Kong	Hong Kong SAR, China
Aaditya Pratap Sanyal	Assistant Professor, Department of Architecture, Planning and Design, IIT (BHU), Varanasi	India
Balaji Kalluri	Assistant Professor, Environmental Studies, School of Liberal Education, FLAME University, Pune	India
Deepa Rani R	Associate Professor, Department of Architecture and Planning, College of Engineering, Thiruvananthapuram	India
Hina Zia	Dean, Faculty of Architecture and Ekistics | Head, Department of Planning and Department of Design & Innovation, Jamia Millia Islamia, New Delhi	India
Jay Dhariwal	Assistant Professor, Department of Design, IIT Delhi	India
Jayasree T K	Assistant Professor, Department of Architecture and Planning, NIT, Calicut	India
Krishna Kumar Dhote	Professor, Architecture and Planning, Maulana Azad National Institute of Technology Bhopal	India
Krishan Upadhyay	Assistant Professor, Institute of Architecture and Planning, Nirma University, Ahmedabad	India
Krishnendu Sinha	Convener Ventilation Committee and Professor of Aerospace Engineering, IIT Bombay	India
Manoranjan Sahu	Associate Professor, Environmental Science and Engineering Department; Associate Faculty, Centre for Machine Intelligence and Data Science; Associate Faculty, Interdisciplinary Programme in Climate Studies, IIT, Bombay	India
Nidhi Gupta	Associate Director - Policy, R&D, Program Design and Implementation Environmental Design Solutions (EDS), New Delhi	India
Nitin Naik	Managing Partner, Dew Point, Goa	India
Piyush Varma	Associate Director, Technology & Software, Environmental Design Solutions (EDS)	India
Prashant Anand	Assistant Professor, Department of Architecture and Regional Planning, IIT Kharagpur	India
Praveen Kumar Yadav	Senior Project Manager - Auditing, Testing and Commissioning Environmental Design Solutions (EDS)	India
Priyanka Kulshreshtha	Asst. Professor, University of Delhi, and Co-Founder, Society for Indoor Environment (SIE)	India
Raja Singh	Post Doctoral Fellow, Drexel Dornsife School of Public Health and Visiting Faculty at School of Planning and Architecture, New Delhi	India
Rajan Rawal	Professor, CEPT University and Senior Advisor, Centre for Advanced Research in Building Science and Energy (CARBSE), CEPT Research and Development Foundation (CRDF), CEPT University, Ahmedabad	India
Rajyalakshmi Garaga	Assistant Professor, Centre for Emerging Technologies for Sustainable Development (CETSD), IIT Jodhpur	India
Sameer Patel	Assistant Professor, Civil Engineering and Chemical Engineering, IIT Gandhinagar	India
Satish Kaipilyawar	Associate Project Director, Society for Health Allied Research & Education-India, Hyderabad	India
Shailee Goswami	Tech Lead - Building Science, Saint Gobain Research	India

Shailendra K. Mandal	Associate Professor, Department of Architecture and Planning, NIT, Patna	India
Srinidhi Balasubramanian	Assistant Professor, Environmental Science & Engineering Department | Affiliate Faculty, Centre for Climate Studies, IIT Bombay	India
Subbaiyan G	Professor, Department of Architecture, NIT, Tiruchirappalli	India
Sukanya R	Consultant - Clinical Microbiology, Infection Control and Prevention, Infectious Diseases, Quality Accreditation & Training	India
Venu Shree	Assistant Professor (BHAVAN Fellow at University of Wisconsin, Milwaukee), Department of Architecture, NIT Hamirpur	India
Yash Shukla	Principal Researcher and Centre Head, Centre for Advanced Research in Building Science and Energy (CARBSE), CEPT Research and Development Foundation (CRDF), CEPT University, Ahmedabad	India
Asit Kumar Mishra	Marie Skłodowska-Curie fellow, School of Public Health, University College Cork	Ireland
Hom Bahadur Rijal	Research Professor and Dean, Graduate School of Environmental and Information Studies, Tokyo City University	Japan
Sushil B. Bajracharya	Assistant Dean, Institute of Engineering (IOE) & Professor, Department of Architecture, IOE, Tribhuvan University, Kathmandu	Nepal
Priya Vishnu	Lecturer, School of Built Environment, Massey University, Auckland	New Zealand
Mohamed Hamdy	Full Professor, Building Performance Simulation and Optimization, Department of Civil and Environmental, Norwegian University of Science and Technology	Norway
Chandra Sekhar	Professor and Program Director MSc (Building Perf & Sustainability), Department of the Built Environment, College of Design and Engineering, National University of Singapore	Singapore
Tham Kwok Wai	Associate Professor, Department of the Built Environment, College of Design and Engineering, National University of Singapore	Singapore
Indrika Rajapaksha	Professor, Department of Architecture, University of Moratuwa	Sri Lanka
Dusan Licina	Assistant Professor, Director of the Human-Oriented Built Environment Lab (HOBEL), École polytechnique fédérale de Lausanne (EPFL)	Switzerland
Arthur Frank	Professor, School of Public Health, and Professor, College of Medicine, Drexel University, Philadelphia	USA
Edward A, Nardell	Professor of Medicine, Harvard Medical School, Massachusetts	USA
Gary Adamkiewicz	Associate Professor of Environmental Health and Exposure Disparities, Department of Environmental Health, Harvard T.H. Chan School of Public Health, Boston	USA
Catherine Noakes	Professor of Environmental Engineering for Buildings and Deputy Director of Leeds Institute for Fluid Dynamics, University of Leeds	United Kingdom
Prashant Kumar	Professor & Chair in Air Quality and Health; Co-Director, Institute for Sustainability, and Founding Director, Global Centre for Clean Air Research, University of Surrey	United Kingdom

Inbound Committee

Nitin Naik	Managing Partner, Dew Point, Goa	India
Priyanka Kulshreshtha	Asst. Professor, University of Delhi, and Co-Founder, Society for Indoor Environment (SIE)	India
Shivraj Dhaka	IGBC	India

Poster and Speaker Support Committee

Aaditya Pratap Sanyal	Assistant Professor, IIT (BHU) Varanasi	India
Ajay Taneja	Professor, Dean Science, Pro Vice Chancellor, DR B R Ambedkar University, AGRA	India
Asit Mishra	Research Fellow, University College, Cork	Ireland
Balaji Kalluri	Asst. Professor, FLAME University, Pune	India
Bharath Kumar Dudam	Assistant Professor, Department of Civil Engineering, Chaitanya Bharathi Institute of Technology, Hyderabad	India
Chunhui Li	Tongji University, Shanghai	China
Elise Huglo	PhD student, UT Austin	USA
Emma Gribben	Head of Office of CEO, iAIR Group, Dublin	Ireland
Harshini G	Green building analyst, Environmental design solutions pvt ltd, Bengaluru	India
Huice Mao	Tongji University, Shanghai	China
Jonathan Griener	B.Eng., Technische Hochschule Nürnberg Georg Simon Ohm, Nueremberg	Germany
Kaiting Chang	Assistant Environment Specialist, Ministry of Environment, Taipei	Taiwan(R.O.C)
Pratima Singh	Associate Professor, Institute of Home Economics, University of Delhi	India
Shailendra Kumar Mandal	Associate Professor (ECR), National Institute of Technology Patna	India
Veerendra Sahu	Postdoctoral Fellow, Indian institute of technology Bombay	India
Venu Shree	Assistant Professor, NIT Hamirpur	India

Registration Committee

Alireza Norouziasas	Research Scholar, Norwegian University of Science and Technology (NTNU)	Norway
Alok Kumar Thakur	Research Scholar, IIT Gandhinagar	India
Ashutosh	Infection Control Nurse, Sarvodaya Hospital	India
Bharath Kumar Dudam	Assistant Professor, Department of Civil Engineering, Chaitanya Bharathi Institute of Technology, Hyderabad	India
Deepa Rani R	Associate Professor, Department of Architecture and Planning, College of Engineering, Trivandrum	India
Himanshu Patel Tuniki	Research Scholar, Kaunas University of Technology (Kaunas, Lithuania)	India

Kaiting Chang	Assistant Environment Specialist, Ministry of Environment	Taiwan (R.O.C)
Kapil Kapoor	Founder - CTO, Vayuguard Climate Tech Private Limited	India
Nishchaya Kumar Mishra	Research Scholar, IIT Gandhinagar	India
Sameer Patel	Assistant Professor, IIT Gandhinagar	India
Saurabh Choudhary	Research Scholar, Central University of Rajasthan, Kishangarh	India
Vandana Jha	Nursing Officer, GTB Hospital	India

Promotional Committee

Adhish Kumar Khadka	Student, Pulchowk Campus	Nepal
Alok Kumar Thakur	PhD Student, IIT Gandhinagar	India
Asit Mishra	Research Fellow, University College Cork	Ireland
Vandana Jha	Nursing Officer, GTB Hospital, New Delhi	India
Veerendra Sahu	Postdoctoral Fellow, IIT Bombay	India

Student Representatives

Prativa Lamsal

Himanshu Patel Tuniki

Keynote Speakers

Arthur Frank

Clinical Professor of Public Health, Professor of Medicine, Drexel University, USA

He is a physician board certified in both Occupational Medicine and Internal Medicine serving as a Clinical Professor of Public Health and a Professor of Medicine at Drexel University in Philadelphia. His PhD is in Biomedical Science for which he studied the biological effects of asbestos on respiratory tissues. He has been an academic physician for almost 50 years and has worked in many countries around the world including India, Sri Lanka, China, Mongolia, Brazil, and elsewhere studying primarily occupational and environmental lung diseases. An author of more than 250 publications, he has been an advisor to many governmental agencies and received numerous awards including a lifetime achievement award from the international Collegium Ramazzini.

Lecture: *Toxins in the Indoor Environment: Contaminants from Workplaces and the Ambient Environment*

Chandra Sekhar

Professor and Programme Director, Dept. of the Built Environment, National University of Singapore

More than three decades as an educator and researcher in the areas of thermal comfort, ventilation and IAQ, airborne infection control, energy-efficient HVAC systems, integrated building design and building energy analysis. Published widely in these areas in international peer-reviewed journals and conferences. Recognised through several awards from ASHRAE, SHASE Japan, ASEAN, Singapore and India, notable ones among which include ASHRAE (2024 Louise and Bill Holladay Distinguished Fellow Award, 2021 E. K. Campbell Award of Merit, 2014 Environmental Health Award, 2013 Exceptional Service Award), SHASE (2019 Uichi Inouyi Memorial Asian International Award), India (2023 MNIT Life Time Achievement Award) and Singapore (2004 The Enterprise Challenge Award).

Lecture*: Mastering IAQ: The 5Ws and H Decoded*

Dusan Licina

Assistant Professor, Director of HOBEL, EPFL, Switzerland

He brings 15 years of expertise in advancing sustainable and healthy indoor environments through research on air pollutant dynamics, exposure science, and ventilation. He is a recipient of the ASHRAE Ralph G. Nevins Award and the ISIAQ Yaglou Award.

Lecture*: Continuous Air Quality Sensing: The Next Frontier of the Built Environment*

Marcel Schweiker

Professor, Head of Chair of Healthy Living Spaces, RWTH Aachen University, Germany

Nearly two decades in research on the interaction between occupants and buildings related to perception, behaviour, health and well-being in a multi-domain context combining thermal, visual, IAQ and acoustics. Member of international and national standardization bodies like DIN and ISO and international working groups under the umbrella of IEA EBC.

Lecture*: Bridging Domains: Emphasizing Human Needs in the Quest for Human-Building-Resilience*

Marianne Touchie

Director, BEIE Lab, University of Toronto, Canada

Her interdisciplinary research program explores the interactions between occupant behaviour, the building enclosure and mechanical systems and how these three factors can be engineered to improve energy performance, indoor environmental quality and occupant wellbeing, including comfort and health. She holds a Canada Research Chair (Tier 2) in Sustainable Urban Housing, is one of Canada's inaugural Clean 50 and has won numerous early career research awards including the ASHRAE New Investigator Award.

Lecture*: Beyond Health & Comfort: Exploring retrofit impacts on resident wellbeing*

Pawel Wargocki

Professor, Technical University of Denmark

He is an ISIAQ, REHVA, and ASHRAE Fellow. He is an academic teacher, scientist, industry consultant, and IAQ expert. He is a mechanical engineer by education. He has expertise in environmental psychology, physiology, and exposure monitoring and more than 25 years of experience in research on human requirements in indoor environments. He graduated from the Warsaw University of Technology in Poland and received his Ph.D. from the Technical University of Denmark, where he has been teaching and performing research ever since. He is known for his seminal work demonstrating that poor indoor environmental quality affects the performance of office work and learning. This and other research work influenced requirements for ventilation and air cleaning.

Lecture: *Demystifying Healthy Buildings: A Call for a Wider Practical Application and Affordability*

Richard de Dear

Professor and Director, Indoor Environmental Quality Laboratory, University of Sydney

Over the last 35 years, Professor Richard de Dear has focused his research career on defining what occupants want and need from their built environments, and assessing the performance of buildings in terms of meeting those requirements. He is currently the most highly cited living researcher in the domain of thermal comfort, with over 250 peer-reviewed papers plus several monographs on the subject. Within that body of research it is his adaptive model of thermal comfort that's had the greatest impact, not just on the research community but also on the design and operation of actual buildings. De Dear's adaptive model underpins the American Society of Heating and the Refrigerating and Air Conditioning Engineers' thermal comfort standard, ASHRAE 55-2004, 2010, 2013, which in turn, informs several national thermal comfort standards around the world.

Lecture: *Cognitive performance across adaptive comfort zone*

International Society of Indoor Air Quality and Climate (ISIAQ)

ISIAQ was founded in 1992 by 109 international scientists and practitioners following the 5th International Conference on Indoor Air Quality and Climate, Indoor Air '90 (Toronto, 1990). It is an international, independent, multidisciplinary, scientific, non-profit organization whose purpose is to support the creation of healthy, comfortable and productivity-encouraging indoor environments.

As a Society, our main objective is to actively promote international and interdisciplinary communication and the exchange of valuable information pertaining to the indoor environment, indoor air, climate, science, and health. We do so by diligently publishing and encouraging the publication of research and studies in the field. Moreover, we take great pride in organizing, sponsoring, and supporting various initiatives like meetings, conferences, and seminars, all dedicated to shedding light on the importance of indoor air quality and climate. In addition, we remain committed to developing, adapting, and maintaining guidelines that serve as a blueprint for enhancing indoor air quality and ensuring a healthy indoor environment. Furthermore, our dedication to this cause extends to our collaboration with government bodies, agencies, and other societies that share a common interest in the well-being of the indoor environment and the impact it has on human health. By working together, we strive to create a cleaner and healthier indoor environment for all.

Members of the Society are:

- Scientists involved in all aspects of indoor air quality
- Government and regulatory professionals
- Medical practitioners
- Occupational health professionals
- Building owners and managers
- Building, construction and air-conditioning engineers
- Architects
- Environmental lawyers

Some of ISIAQ's most important activities include:

- Publication of a Newsletter which carries news and information.
- Developing guideline documents by a number of STCs focused on specific issues. STCs convene during ISIAQ conferences.
- Organizing Indoor Air and Healthy Buildings series as primary Society conferences.

If you are interested in becoming a member of ISIAQ, please visit: isiaq.org

ISIAQ INTERNATIONAL SOCIETY OF INDOOR AIR QUALITY AND CLIMATE

Contents

Investigation of Size-Segregated Particle Penetration and Infiltration Dynamics in an Analytical Laboratories

Prashant Nawale *, **Mohd Zaid, Manoranjan Sahu**
Indian Institute of Technology Bombay Mumbai 400076, India.

Exposure to toxic components in indoor particulate matter (PM) is linked to various adverse health effects, particularly respiratory and cardiovascular conditions. A considerable portion of human exposure to outdoor-origin PM occurs indoors, especially in institutional analytical laboratories. This study focuses on indoor and outdoor PM concentration and size distribution in analytical laboratories in Mumbai, to understand the infiltration dynamics of outdoor particles. The objective is to investigate penetration and infiltration dynamics of size-segregated PM in a laboratory setting. This study was conducted under both occupied and unoccupied conditions, utilising portable instruments to measure particle sizes ranging from 0.007 to 10 μm. The study reveals that indoor aerosols exhibit less variability in size distribution than outdoor aerosols, depending on indoor conditions. Indoor-to-outdoor (I/O) ratios for ultrafine particles (UFPs) number concentrations are highly dependent on particle size. The mean penetration factors for UFPs, fine and coarse were 0.72 ± 0.13, 0.6 ± 0.1 and 0.54 ± 0.09, respectively, with relative uncertainties below 20%. At air exchange rates of 0.67–3.58 hr^{-1}, fine particles exhibited the highest I/O ratios (0.67–0.98). The mean fine-to-coarse particle ratios were 0.88 ± 0.06 indoors and 0.78 ± 0.11 outdoors. Comparable size-segregated particle number distributions for UFPs in indoor and outdoor settings highlight the combined impact of indoor activities and outdoor PM infiltration. These results highlight the significant infiltration of ultrafine particles into indoor environments and their potential health implications. This research provides valuable insights into the behaviour of PM infiltration in institutional settings and its impact on indoor air quality.

Keyword: *Penetration factor, Particulate Matter, Particle Size Distribution*

Multizonal Modeling in a Residential Apartment Using Physics Informed Long Short-Term Memory Approach

Alok Kumar Thakur * **, Sameer Patel**
Indian Institute of Technology Gandhinagar, India

Most indoor air pollution research focusing on modeling and material balance assumes well-mixed conditions, which is usually not the case, especially in bigger and multizonal spaces. Studying the inter-zonal transport of pollutants and their governing factors provides critical insights into the fate and transport of pollutants from emission zones to different zones in a multizonal indoor space. The current work focuses on predicting pollutant concentrations in different zones of a residential apartment using pollutant levels measured by a monitor in one zone. The governing parameters here include inter-zonal flow and deposition rates. The work consists of two parts - (a) estimation of inter-zonal flow rates between different zones and (b) spatio-temporal prediction of PM concentrations using a deep learning model.

Spatio-temporal measurements of $PM_{2.5}$ and CO_2 were conducted in different zones of a multizonal apartment for different scenarios. Inter-zonal flow rates, estimated via a custom MATLAB algorithm and experimentally derived deposition rates, were input parameters for a physics-informed long short-term memory (PI-LSTM) model. Unlike conventional LSTM models, the PI-LSTM framework incorporates physics-based constraints and governing equations, allowing for accurate predictions even with limited training data. Validation was performed using experimentally obtained datasets, with the normalized mean bias serving as a key performance metric for evaluating the predictive capabilities of both LSTM and PI-LSTM models.

Keyword: *Indoor air quality, Multizonal modeling, Physics-informed deep learning*

INFLUENCE OF ASPECT RATIO AND THE HEIGHT OF THE BUILT SPACE ON IAQ AND TEMPERATURE DISTRIBUTION WITHIN THE INSTITUTIONAL KITCHEN ENVIRONMENT

Jesna Mathew * , **Subbaiyan G**

Department of Architecture, National Institute of Technology, Tiruchirappalli

Indoor air quality (IAQ) and thermal comfort are two critical aspects influencing the health and productivity of occupants in the workplace. Ventilation plays an important role in maintaining thermal comfort and achieving optimal indoor air quality (IAQ) in the built environments. The IAQ and thermal environment of institutional kitchens are critical for the health and productivity of workers, as these spaces are characterized by significant heat generation and pollutant emissions due to cooking activities. The dispersion of pollutants and heat within the kitchen can be mitigated through good ventilation design.

This study examines the influence of the geometry of kitchen space (in terms of aspect ratio and height) on IAQ and temperature distribution within institutional kitchens. The ventilation effectiveness is used as a parameter to analyse IAQ. Various geometric configurations of kitchen spaces were analysed through computational fluid dynamics (CFD) simulations and experimental validation to assess airflow patterns, ventilation effectiveness and temperature distribution. Ansys Fluent is used for CFD simulation to various geometric configurations.

The results indicate that a reduced aspect ratio enhances the IAQ and reduces the temperature of the kitchen space to that of the outdoor air temperature by efficiently removing the high heat generated by the burners. Similarly, IAQ improved with increased roof height. These findings highlight the importance of spatial geometry in designing naturally ventilated kitchens with enhanced IAQ and thermal performance, offering practical recommendations for the design community to create healthier and more efficient workspaces.

Keyword: *IAQ, Ventilation, Aspect Ratio*

THE STUDY ON INFLUENCE OF ASPECT RATIO AND HEIGHT OF THE BUILT SPACE ON IAQ AND TEMPERATURE DISTRIBUTION WITHIN THE INSTITUTIONAL KITCHEN ENVIRONMENT.

Jesna Mathew[1], * Gnanasambandam Subbaiyan[1],

1 National Institute of Technology Tiruchirappalli, India * 401120002@nitt.edu

SUMMARY

Indoor Air Quality (IAQ) and thermal environment of institutional kitchens are critical for the health and productivity of workers, as these spaces are characterized by significant heat generation and pollutant emissions due to cooking activities. Effective ventilation design can mitigate heat and pollutant accumulation. This study examines the influence of the geometry of kitchen space, specifically aspect ratio and ceiling height, on IAQ and temperature distribution. The validated computational fluid dynamics (CFD) simulations using ANSYS Fluent were used to analyse the influence of various geometric configurations on airflow patterns, IAQ and temperature distribution. The results indicate that a reduced aspect ratio enhances the IAQ and air velocity. Similarly, air velocity at the cooking zone decreased with increased roof height. The findings highlight the importance of spatial geometry in designing naturally ventilated kitchens with enhanced IAQ and thermal performance, offering practical recommendations for the design community to create healthier and energy-efficient kitchen environments.

KEYWORDS

IAQ, Temperature, Aspect Ratio, Age of air, Ventilation, CFD

1. INTRODUCTION

Poor Indoor air quality and thermal comfort in the workplace have detrimental effects on the health and well-being of occupants. The commercial kitchen spaces, where many people work as employees, are particularly vulnerable to high temperatures and air pollution caused by cooking activities. Consequently, IAQ in commercial kitchens is a significant occupational health concern that neds to be addressed. Previous studies done in Kitchen environments have reported elevated levels of pollutants such as polycyclic aromatic hydrocarbons, carbon monoxide, carbon dioxide and particulate matter (Singh et al., 2016). The health risks of kitchen air pollution range from acute and chronic respiratory symptoms to potential carcinogenic risks (Saha et al., 2012). Cooking in a commercial kitchen is more intense and demanding in comparison to a residential kitchen and can lead to instantaneous cooking pollution from volatile organic chemicals, posing a health hazard to cooks (Durban, 2022).

IAQ in the kitchen environments is affected by various factors, such as the fuel type, kitchen design, the type of dishes cooking and ventilation strategies (Tsang et al., 2023). The mitigation measures, such as range hood and well-designed ventilation systems, can significantly reduce exposure to cooking related pollutants (Tsang et al., 2023).

Institutional kitchens represent a specific category of commercial kitchens designed to cater to a particular community. In developing countries such as India, many institutional kitchens lack adequate mechanical ventilation systems to extract heat and pollutants from cooking activity. As a result, these workplaces are especially vulnerable to poor air quality and high accumulation of heat, which lead to various kinds of health problems among workers. The literature has also pointed towards the necessity for proper ventilation in different kitchen settings, including residential kitchens, commercial kitchens and railway pantry cars (Alam & Salve, 2021; Chen et al., 2025). In addition, studies indicate that a well-designed mechanical ventilation system is effective at removing cooking pollutants (Tsang et al., 2023), while natural ventilation can also be impactful when properly designed (Lee et al., 2012). The building's aspect ratio and opening sizes are key variables that influence the airflow in a naturally ventilated building (Debnath et al., 2016; Mathew & Subbaiyan, 2024). The World Health Organization (WHO) also says that improvements in building factors, such as optimized opening-to-wall ratios, aspect ratios, improved cookstoves, and enhanced airflow in the cooking zone are important for a healthy kitchen (World Health Organization, 2014).

The current article aims to investigate the kitchen space geometry in terms of aspect ratio and ceiling height effects on indoor air and temperature distribution and IAQ in a naturally ventilated institutional kitchen. Ventilation effectiveness and indoor air quality were assessed using air velocity and age of air as key parameters, particularly within the cooking zone. A full-scale experiment and measurements were conducted in situ, replicating real operating conditions, which were then used to model and validate CFD model. This validated simulation was then used to explore the impact of variations in kitchen geometry on characteristics of airflow and thermal distribution inside the kitchen space.

2. METHODOLOGY

2.1. Full-scale experimental design

An institutional kitchen (hostel mess kitchen) in a large campus situated in Tiruchirappalli, Tamil Nadu, India (10.7905^0 N, 78.7047^0 E) is selected for the full-scale experimental study (Figure 1). This kitchen is a naturally ventilated kitchen having an area of 148.8 m^2 and serving around 600 students. The experiment was carried out during peak summer conditions in Tiruchirappalli over a three-day period (June 15–June 17) in fully natural ventilation mode. During typical culinary activities, airspeed and temperature were monitored at multiple locations (see figure 1) within the kitchen.

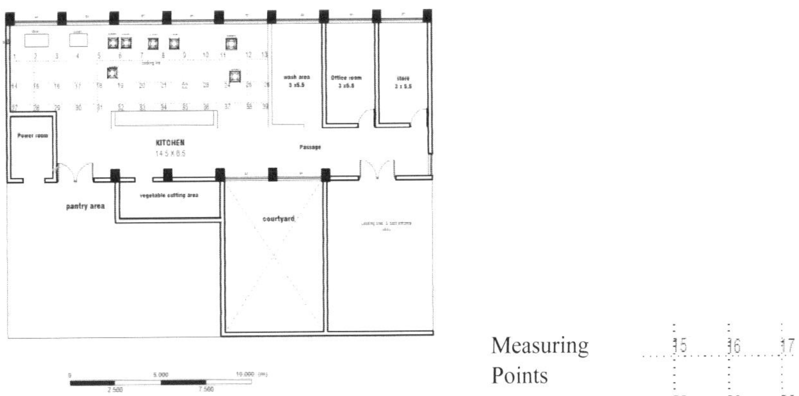

Figure 1. Plan of the kitchen- experimental case study

2.2 CFD simulation - Numerical methodology

The literature study points towards the fact that utilising a validated model, primarily a CFD Model, is the standard method to characterize the implications of different designs on natural ventilation in indoor environments. (Izadyar et al., 2020, 2023; Omrani et al., 2015). ANSYS Fluent 2023 R1 was deployed to test the effects of possible configurations on indoor airflow and temperature in the indoor area using 3D RANS model coupled with standard k–ε turbulence model. ANSYS Discovery Space Claim was used to model the three-dimensional geometry of the kitchens. A larger domain was created around the case for the airflow analysis. The outer domain size and extent were decided as recommended by Izadyar et al. (2020). The height of the building (H) serves as the base for the dimensions of the larger domain. The upstream, downstream, lateral, and height of the larger domain are 3H, 15H, 3H, 3H, and 6H, respectively. An unstructured grid with tetrahedral volume was generated for the numerical study. A finer and refined grid was used for the building surfaces and openings, and the most refined grid was used for the burner surface, which is .001. A wind velocity of 1.67 m/s measured by a local weather station installed on the site at a height of 12m above the ground is considered as the inlet boundary condition of the larger outer domain. Weibull function (1) was used to apply the Inlet Velocity.

$$V_h = \mathrm{Vr}\left(\frac{\mathrm{h}}{\mathrm{hr}}\right)^{\alpha} \quad \alpha\text{=0.27 } \textit{is the terrain} \text{ roughness for suburban area} \quad (1)\ \text{(Izadyar et al., 2023)}$$

where V_h is the velocity inlet profile related to a specific height, and Vr and hr represent the velocity and height at the wind station, as the simulation reference. The outdoor air temperature is specified as 35.2 ^{0}C, whereas the surface temperatures for the floor, sidewalls and ceiling are considered as 32 ^{0}C, 33 ^{0}C and 34 ^{0}C respectively, as per the measurements conducted on the experimental study. The burner temperature is given as 1240 K (Saha et al., 2012). The coupled algorithm was used for pressure-velocity coupling.

2.3. Validation of CFD simulation

In-situ measurements were used to validate the CFD Model. For this, the velocity and temperature measured near the location are compared with the CFD simulation for the same kitchen settings created in the computer domain. The data is shown in Figure 2. The RMSE (Root Mean Square Error) percentage were calculated. The RMSE percentages for velocity are 16.1%, 20.03%, and 12.4%, and for temperature are 17.3%, 6.87%, and 2.09% at the 0.1 m, 1.1 m, and 1.6 m planes, respectively.

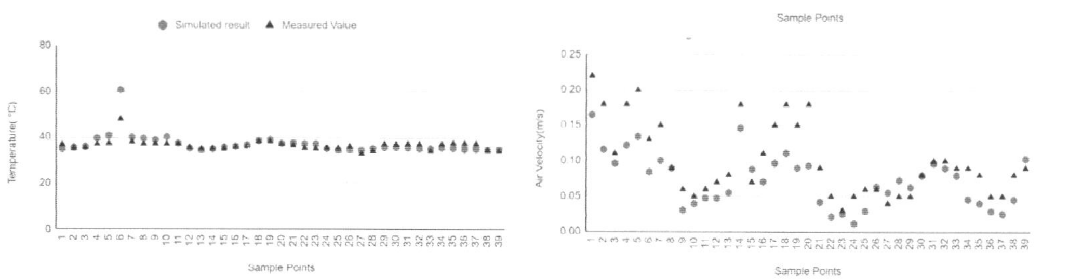

Figure 2. Experimental and simulated values for air temperature and air velocity for level 1.1m

2.3 Configuration test

The kitchen for the simulation study was modelled in Ansys with the same boundary conditions and spatial dimensions observed in situ. This approach helped to accurately validate the CFD model against field measurements, setting a reliable base case for further parametric investigations. Five different aspect ratios, ranging from 1.0 to 0.5, were examined and are labelled as A1 through A5 (as illustrated in Figure 3), with A5 corresponding to the configuration that closely matches the experimental aspect ratio (1:0.48). To evaluate the influence of room height, three different values were considered: 4.0 m, 4.5 m, and 5.0 m. A total of 15 cases were considered. In all configuration tests, the cooking line was positioned near the leeward wall. This placement was selected due to its improved performance in reducing temperature spread and facilitating pollutant dispersion when compared to the windward-side placement used in the experimental setup. Seven burners were included in the model, based on observations from the experimental study. The direction of the wind was maintained perpendicular to the longer wall, similar to the experimental arrangement. The window-to-wall ratio (WWR) was kept constant at 26 % of the floor area and was distributed equally on both the leeward and windward walls across all the test cases, ensuring that the effects of aspect ratio and height on indoor air distribution could be examined independently of window geometry variations. The outdoor wind velocity considered for configuration testis 1.67 m/s.

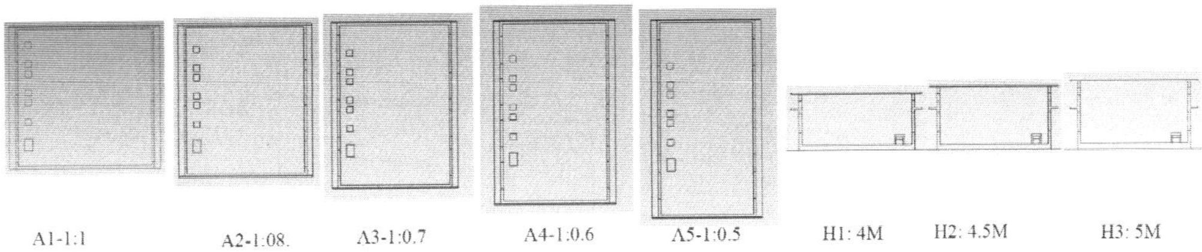

Figure 3. Configuration test- Aspect ratio and Height variation considered

3. RESULT AND DISCUSSION

The indoor airflow pattern, temperature distribution, and Local Mean Age of Air (LMA) were compared across various configurations (see figures 4 and 5). Average air velocity, temperature, and age of air in the cooking zone are compared in Figure 6. Among all configurations, A1H1 exhibits the highest velocity in the cooking zone, which decreases progressively from A1H1 to A4H3. Interestingly, in A5H3, a very slight increase in velocity is observed compared to A4H3 and A5H2, likely due to a localized low-pressure zone above the cooking line, which helps to increase more airflow toward the cooking zone as seen in Figure 4, though it remains lower than in A5H1. It is observed that as the aspect ratio increases, the airflow at the cooking zone reduces, leading to increased age of air. Similarly, increasing the height (floor area same) results in lower airflow and higher age of air in the cooking zone.

As seen in Figure 5, the temperatures experienced in the cooking zone remain close to the inlet air temperature of 35.2 °C, with minimal variation. This indicates that effective airflow in the cooking zone helps maintain thermal conditions near the inlet level or slightly lower, thereby preventing excessive heat build-up from burners and limiting heat spread to the surrounding cooking area.

The age of air indicates air quality inside the space. The LMA at the breathing plane is shown in Figure 5. As the aspect ratio increases from 1:1 to 1:0.5, the LMA increases, as reflected by the transition to greener shades. The A5H3 configuration, which had the largest aspect ratio and maximum height, showed the highest mean age of air (area averaged mean age of air at breathing plane) about 32.71s. Especially, the mean age of air at the breathing level increased by over 93.45% from configuration A1H1 to A5H2, suggesting significantly lower ventilation effectiveness. This implies that the A1H1 model offers better air distribution.

An analysis of Figure 4, which depicts air distribution within the kitchen space, reveals that the formation of low airflow zones (indicated by blue areas) increases with the aspect ratio. This occurs because the window-to-wall ratio (WWR) remains constant and is evenly distributed between the windward and leeward walls across all aspect ratio models. However, as the lengths of the windward and leeward walls increase, the spacing between adjacent windows also becomes larger. This increased spacing contributes to the development of low airflow zones, indicating a reduction in ventilation effectiveness.

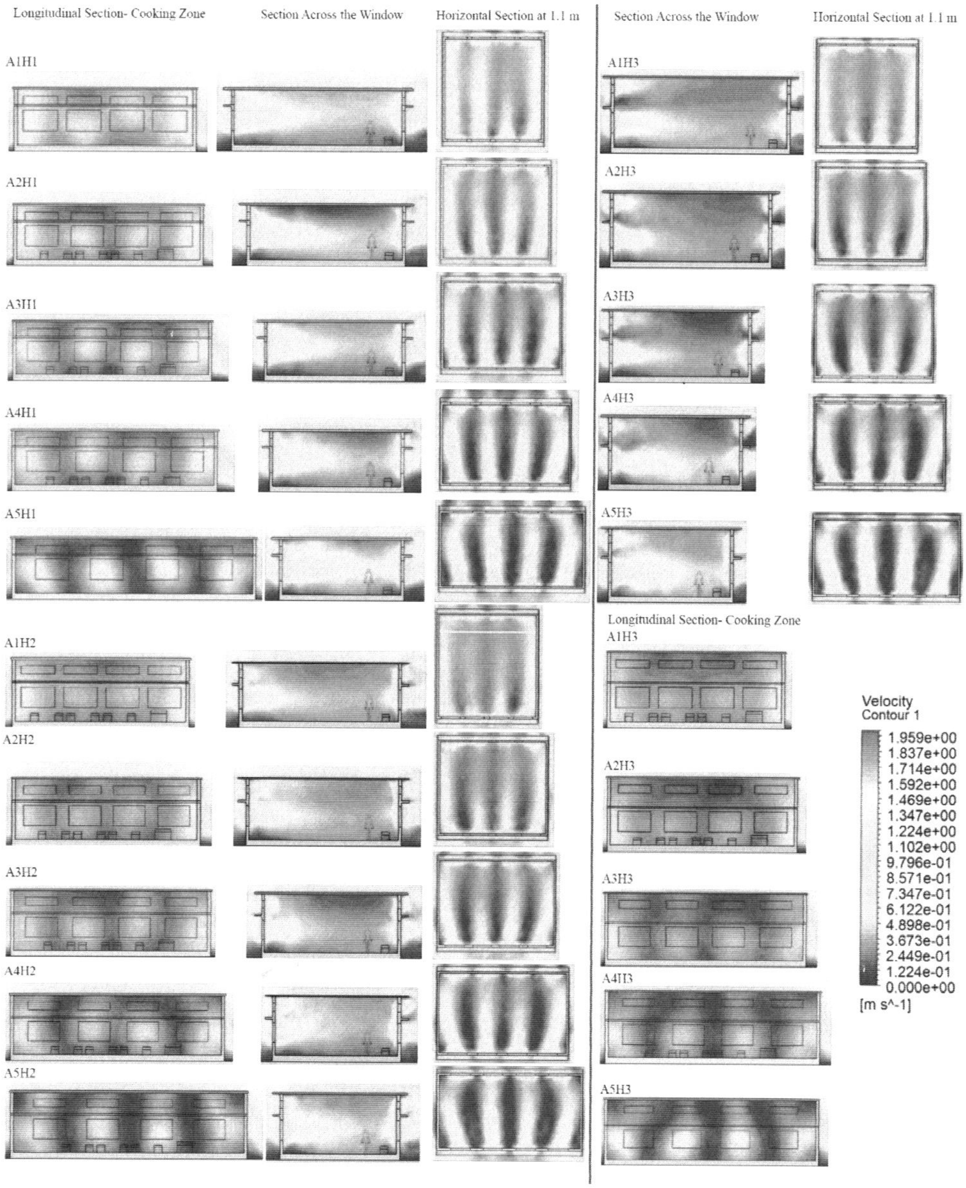

Figure 4. Comparison of air velocity at different planes in different design configurations.

The relationship between aspect ratio and variables (air velocity, air temperature and age of air) for different height considered is shown in figure 7. The results suggest that, with increasing aspect ratio, air velocity in the cooking zone decreases and the mean age of air at the breathing level and cooking zone increases, as reported in an earlier study by Mathew J. and Subbaiyan

(2024). Furthermore, the CFD analysis images indicate that air distribution within the space is influenced by the arrangement and distribution of inlets and outlets along the windward and leeward walls. Additionally, increasing the room height alone does not significantly improve air distribution or reduce the air age in the cooking zone or at the breathing level. Configuration tests involving various combinations of inlet and outlet arrangements, as well as varying opening areas, may add further clarity to the influence of building parameters on ventilation performance and hence inform the selection of optimized kitchen design for improved indoor air quality.

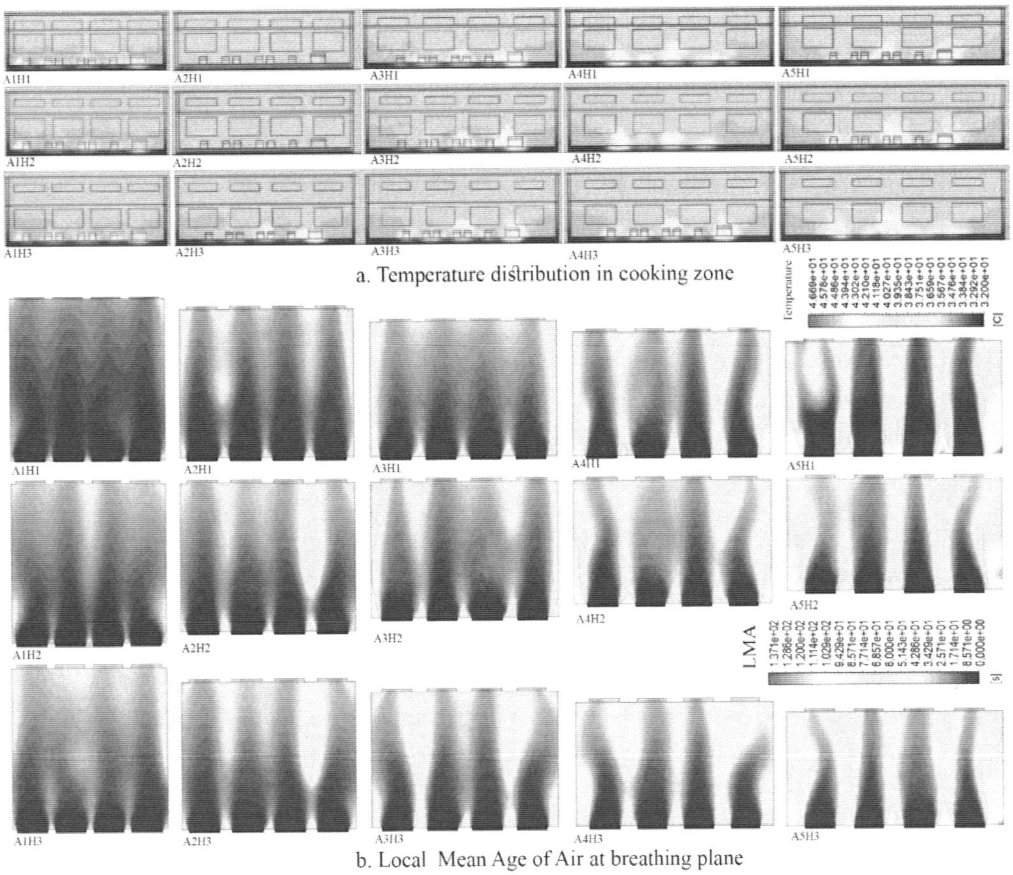

Figure 5. a) Comparison of Air temperature at cooking zone (longitudinal section). b) Comparison of LMA at breathing plane (1.6m) in different design configurations

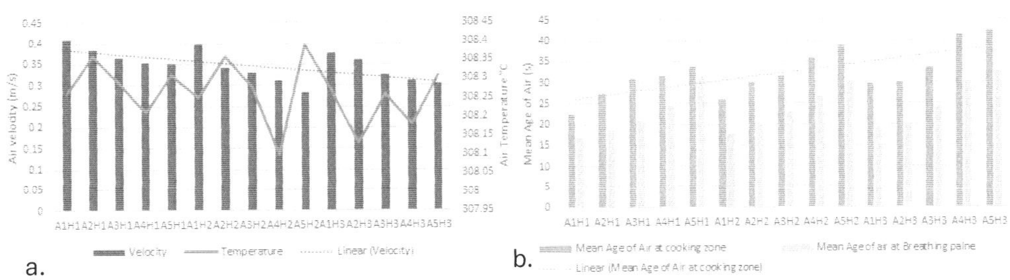

Figure 6. a) Average air velocity and air temperature at cooking zone b) MAA at cooking zone and breathing plane.

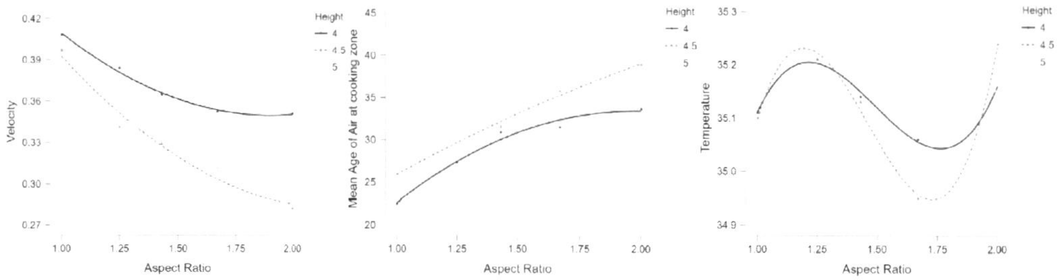

Figure 7: Relationship between Aspect ratio and Variables (Air velocity, age of air and air temperature) across different heights.

4. CONCLUSION

The present study used a validated CFD model to simulate airflow, temperature, and LMA under various configuration scenarios. The results indicate that decreasing the aspect ratio from 1:0.5 to 1:1 enhances the air velocity in the cooking zone and decreases the mean age of air at the breathing plane, signifying better ventilation effectiveness and indoor air quality. Similarly, increasing height increases the mean age of air at the cooking zone and breathing plane and decreases air velocity at the cooking zone. The results emphasize the importance of the spatial proportions of kitchen space on natural ventilation effectiveness. Among the tested configurations, the A1H1 model, with a lower aspect ratio and lower height, demonstrated the most favourable performance. Further study to look at the impact of inlet and outlet opening arrangement, varying opening area and varying cooking line arrangement on ventilation effectiveness, IAQ and temperature distribution in kitchen space may be helpful to find an optimal design solution for kitchen design.

5. ACKNOWLEDGEMENTS

The authors gratefully acknowledge the financial support provided by the Prime Minister's Research Fellows (PMRF) scheme, Ministry of Education, Government of India.

6. REFERENCES

Alam, M. S., & Salve, U. R. (2021). Enhancement of Thermal Comfort Inside the Kitchen of Non-Airconditioned Railway Pantry Car. International Journal of Heat and Technology, 39(1), 275–291.

Chen, S.-W., Chang, C.-Y., & Chan, Y.-C. (2025). Assessing the impact of kitchen hoods and ventilation openings on indoor air quality post-cooking. Journal of Building Engineering, 100, 111721.

Debnath, R., Bardhan, R., & Banerjee, R. (2016). Investigating the age of air in rural Indian kitchens for sustainable built-environment design. Journal of Building Engineering, 7, 320–333.

Durban, S. G. (2022). A Review of Food Toxicity and Kitchen Air Pollution: Cooking Food in the Hospitality Industry. Global Conference on Business and Social Sciences Proceeding, 13(1), 1–1.

Izadyar, N., Miller, W., Rismanchi, B., & Garcia-Hansen, V. (2020). Numerical simulation of single-sided natural ventilation: Impacts of balconies opening and depth scale on indoor environment. IOP Conference Series: Earth and Environmental Science, 463(1), 012037.

Izadyar, N., Miller, W., Rismanchi, B., Garcia-Hansen, V., & Matour, S. (2023). Balcony design and surrounding constructions effects on natural ventilation performance and thermal comfort using CFD simulation: A case study. Journal of Building Performance Simulation.

Lee, H., Lee, Y. J., Park, S. Y., Kim, Y. W., & Lee, Y. (2012). The Improvement of Ventilation Behaviours in Kitchens of Residential Buildings. Indoor and Built Environment, 21(1), 48-61.

Mathew, J., & Subbaiyan, G. (2024). Investigation of Ventilation Performance to Improve the Indoor Air Quality of Institutional Kitchens. Journal of Daylighting, 11(1), 39–54.
Omrani, S., Capra, B., Garcia-Hansen, V., & Drogemuller, R. (2015). Investigation of the effect of balconies on natural ventilation of dwellings in high-rise residential buildings in subtropical climate.
Saha, S., Guha, A., & Roy, S. (2012). Experimental and computational investigation of indoor air quality inside several community kitchens in a large campus. Building and Environment, 52, 177–190.
Singh, A., Kamal, R., Mudiam, M. K. R., Gupta, M. K., Satyanarayana, G. N. V., Bihari, V., Shukla, N., Khan, A. H., & Kesavachandran, C. N. (2016). Heat and PAHs Emissions in Indoor Kitchen Air and Its Impact on Kidney Dysfunctions among Kitchen Workers in Lucknow, North India. PloS One, 11(2).
Tsang, T. W., Wong, L. T., Mui, K. W., & Poon, C. Y. (2023). Influences of home kitchen designs on indoor air quality. Indoor and Built Environment, 32(7), 1429–1438.
World Health Organization (Ed.). (2014). WHO guidelines for indoor air quality: Household fuel combustion. World Health Organization.

Distributional Mapping of PM2.5 and CO2 in a Multizonal Apartment Using Spatio-Temporal (Graph Convolution Network and Graph Attention Network)

Alok Kumar Thakur *, **Nishchaya Kumar Mishra, Sameer Patel**
Indian Institute of Technology Gandhinagar, India

Modeling studies on indoor air quality for source and space characterization are prevalent for single-zone environments, primarily due to the simplicity of implementing single-box models under the assumption of well-mixed conditions. However, most urban indoor environments are multizonal, necessitating a multizonal representation for a more comprehensive understanding. Investigating inter-zonal pollutant transport in such spaces is critical to modeling the dynamics of pollutant dispersion from emission sources to other zones and developing effective control strategies. The current study uses spatial interpolation techniques, graph convolutional networks, and graph attention networks to generate high-resolution pollutant concentration maps for a multizonal apartment. Spatio-temporal measurements of $PM_{2.5}$ and CO_2 were conducted across multiple zones of a three-bedroom apartment under different internal door configurations. Conventional surrogate models often struggle to accurately predict spatio-temporal dynamics, particularly in complex multizonal indoor environments. In contrast, advanced graph neural network (GNN) architectures, such as spatio-temporal graph convolutional networks (ST-GCNs) and graph attention networks (GATs), offer a robust framework for generating high-resolution spatio-temporal pollutant distribution maps. These models represent zones and sensor locations as nodes, with edges capturing inter-zonal relationships, enabling precise predictions even with limited training data. Model validation will be performed using experimentally obtained datasets and assessed through various performance metrics to ensure reliability and accuracy.

Keyword: *Indoor air quality, Multizonal prediction modeling, Graph neural networks*

Identifying Ventilation Strategies for Healthier Classrooms using Spatial Modelling of Indoor Carbon Dioxide Concentrations

Ravinder Arigela * [1], **Sathish Kannan**[1], **Pranjal Jejurkar**[1], **Virendra Sethi**[1], **Krishnendu Sinha**[2], **Srinidhi Balasubramanian**[1,3]

[1] *Environmental Science and Engineering Department, Indian Institute of Technology Bombay, Mumbai, India*

[2] *Department of Aerospace Engineering, Indian Institute of Technology Bombay, Mumbai, India*

[3] *Center for Climate Studies, Indian Institute of Technology Bombay, Mumbai, India*

The COVID-19 pandemic has highlighted the need for effective ventilation in high-occupancy indoor environments, such as classrooms, to mitigate airborne disease transmission. Carbon dioxide (CO_2), a by-product of human respiration, is widely used to assess ventilation performance. While CO_2-based measurements are common, Computational Fluid Dynamics (CFD) offer a more detailed evaluation of ventilation dynamics and disease transmission across various occupancy and mitigation scenarios.

In this study, ANSYS Fluent was used to simulate the spatial variation of CO_2 concentrations in a model university classroom with 60 students under different ventilation conditions. Occupants were modeled as cuboidal geometries with a 12 mm nose diameter, emitting CO_2 at typical sedentary rates. Tracer gas dynamics were captured using the species transport equation, while airflow behavior was characterized using the k-ε turbulence model with enhanced wall functions. Ventilation scenarios observed in live classrooms, including different combinations of open doors/windows and ceiling/exhaust fans, were evaluated. Preliminary results indicated that lower CO_2 concentrations less than ~1000 ppm as recommended by ASTM D6245 existed in Door-Ceiling Fans-Exhaust Fans on scenario due to enhanced air mixing. However, higher CO_2 concentrations (>1000 ppm) were observed at the breathing plane (1.2 m) across all scenarios, highlighting localized breathing plume effects compared to standing (1.6 m), sensor (2.45 m), and near-ceiling (3.2 m) heights.

Ongoing efforts are focused on estimating air change rates and spatial infection risk using CFD. This study addresses the limitations of the well-mixed air assumption and provides valuable insights for optimizing classroom ventilation to reduce disease transmission risks.

Keyword: *Indoor air quality, Classroom ventillation, Infection risk*

Investigating CO2 based Indoor Air Quality of Classroom using Experimental Measurements and Computational Fluid Dynamics Approach

Mahima Pareek, Avnish Shukla *, Sairam Jinnawar, Krishnendu Sinha
Department of Aerospace Engineering, Indian Institute of Technology, Bombay-400076, India

Indoor air quality (IAQ) is a crucial factor in ensuring an effective learning environment within classrooms, where carbon dioxide (CO_2) buildup is an important concern. This study investigates the spatial distribution of CO_2 and its temporal decay by conducting controlled experiments in a life-size ventilation testbed replicating a real classroom and utilizing Computational Fluid Dynamics (CFD) simulations. The primary objective is to develop solutions to improve IAQ by increasing the air exchange rate, sufficient to ventilate the entire classroom within a short break between consecutive classes.

Data was collected utilizing real-time CO_2 sensor deployed at various monitoring points to monitor CO_2 decay with various air exchange rate within the ventilation testbed at IIT Bombay. Findings indicate that CO_2 decay rate varies at different monitored locations. Using time-dependent CFD modeling, airflow visualization was observed, indicating poorly ventilated regions having slower CO_2 decay. Furthermore, the statistical analysis using z-test was performed on CO_2 decay rates. Results emphasize the crucial role of optimum ventilation strategies in mitigating CO_2 exposure and enhancing IAQ of classroom. By integrating experimental data with CFD simulation, the study contributes to the development of evidence-based solutions leading to a conducive learning environment.

The findings of this study enable the development of recommendations to achieve the desired air exchange rates and strategic plans that can be incorporated into real classroom settings to improve IAQ. The insights gained can inform the design of ventilation systems that effectively reduce CO_2 concentrations within short time frames, ensuring better air quality during classroom activities.

References:

Du, B., Reda, I., Licina, D., Kapsis, C., Qi, D., Candanedo, J. A., & Li, T. (2024). Estimating Air Change Rate in Mechanically Ventilated Classrooms Using a Single CO2 Sensor and Automated Data Segmentation. Environmental Science & Technology, 58(42), 18788-18799.

Krawczyk, D. A., Rodero, A., Gładyszewska-Fiedoruk, K., & Gajewski, A. (2016). CO2 concentration in naturally ventilated classrooms located in different climates—Measurements and simulations. Energy and Buildings, 129, 491-498.

Norbäck, D., & Nordström, K. (2008). Sick building syndrome in relation to air exchange rate, CO 2, room temperature and relative air humidity in university computer classrooms: an experimental study. International archives of occupational and environmental health, 82, 21-30.

Keyword: *Indoor Air Quality (IAQ), Carbon dioxide (CO2), Computational Fluid Dynamics (CFD)*

Assessment of CFD Modelling of Ceiling Fan for Predicting Indoor Air Quality and Airborne Infection

Sourabh Nandi *, **Paul Jacob, Lagoon Biswal, Janani Murallidharan, Amit Agrawal, Rajneesh Bhardwaj, Krishnendu Sinha**

Indian Institute of Technology Bombay

Computational Fluid Dynamics (CFD) is routinely used to study indoor ventilation and design strategies for better indoor air quality. The assessment of ventilation using Computational Fluid Dynamics (CFD) for ceiling fans remains limited to its impact on air-flow comfort and not on its impact on indoor air quality. There is a need for a comprehensive numerical evaluation of air quality, bio-aerosol dispersion and infection modelling.[1]

This study employs CFD simulations to analyze different fan modelling approaches, including Pressure Jump, Moving Reference Frame (MRF), and Unsteady Reynolds-Averaged Navier-Stokes (URANS) with Sliding Mesh motion, to assess their effectiveness based on key parameters such as Air Residence Time (ART), Air Change Per Hour (ACH) and Relative Quanta Exposure (RQE).[2] This numerical methodology is applied to an IIT Bombay classroom with four ceiling fans and two exhaust fans, examining airflow patterns and contaminant transport. The study compares modelling accuracy, computational time and resource requirements to determine the most efficient approach. Additionally, a procedure will be proposed to extrapolate the results for similar classrooms with varying configurations of ceiling and exhaust fans.

The findings of this research will provide an accurate method of ceiling fan modelling in computational fluid dynamic simulations that can be used effectively in the design of ventilation systems for realistic scenarios.

References:

S. Omrani, S. Matour, K. Bamdad, and N. Izadyar. Ceiling fans as ventilation assisting devices in buildings: A critical review. Build. Environ., 201:108010, 2021.

Mullick, A., Kumaraswamy, G., Mehra, S., Murallidharan, J., Kumar, V., & Sinha, K. (2023). Is well-mixed model of an indoor space with ceiling fans valid for studying pathogen transmission?. Physics of Fluids, 35(10).

Keyword: *Ventilation CFD, Air-flow comfort, ceiling fan*

Multidirectional wind deflectors in Dublin city- A Real world analysis

Madhavan Vasudevan * [1], **Ratnakishore Velamati**[2], **Francesco Pilla**[3], **John Gallagher**[1], **Aonghus McNabola**[4]

[1] *Trinity College Dublin*
[2] *Amrita Vishwa Vidyapeetham*
[3] *University College Dublin*
[4] *RMIT, Australia*

Roof-level interventions have shown potential in the past to make fresher above roof-level air available at the polluted street level, and even so have their adjustable versions. However, the dynamism of environment seldom allows sufficient time for the passive interventions to adapt well. Therefore, a multi-directional wind deflector (MDWD) system is introduced to at least achieve partial wind deflection for varying ambient wind directions. The design was first evaluated for its performance when 1, 2, 3 and no MDWDs were introduced in an isolated street canyon under parallel (0°), orthogonal (90°) and oblique (45°). The downward push of the parallel winds facilitated by the MDWDs reduced personal exposure in upwind region by a minimum of 25.7% across Floors 1 and 2. While the efficacy was more pronounced for the oblique and orthogonal winds in the only higher floors (4-6) of the leeward façade for central and downwind regions due to stronger streamwise washout and throughout the canyon with a minimum 47.4% reduction across a floor, the windward sides experienced a rise in exposure. The street-canyon like environment of Pearse Street in Dublin, Ireland was considered as a case study for evaluating the MDWDs in a real-world environment. The MDWDs under predominant wind condition based on 2024 meteorological measurements resulted similarly in maximum 16.1% reduction and 19.5% increase across specific leeward and windward target facades. Finally, similarities between theoretical and real street analysis were discussed and guidelines were prescribed for enhanced urban ventilation.

Keyword: *Urban ventilation, wind deflectors, street canyons*

Evaluation of airflow and thermal comfort of a classroom in Warm & Humid climate – A case study of a school in Bhubaneswar

Kiriti Sahoo * [1], **Kasthurba A. K**[2], **Deepthi Bendi**[1]
[1] *Assistant Professor*
[2] *Professor*

Thermal comfort is important in classrooms in which natural ventilation plays a critical role in achieving the same. Thermal comfort is extremely crucial in locations in hot & humid zone. Natural ventilation offers significant and economical solutions in such climatic zones. The study presents thermal comfort of a public school located in Bhubaneswar featuring climate responsiveness. The study encompasses impact of natural ventilation on thermal comfort of the classroom during routine occupancy hours. The methodology involved the review and analysis of design features such as layout, position of windows, doors and ventilators. In order to assess the impact of the design features, a simulation model was developed to understand airflow pattern, DBT, RH and stack pressure. It was observed that, ventilation rate of minimum 6 ach was not achieved in the current classrooms scenario. This impacts thermal comfort, thus becoming non-compliant with NBC 2016. As a measure, interventions in the form of retrofits on wall and roof such as solar chimney, double wall, and high SRI paints on roof are explored. The study experimented this through a simulation model. The results of the experiment indicated that achieved airflow rate and resulting thermal comfort was compliant with NBC 2016 . Further, this paper discusses the exploration of retrofit in classrooms and its impact on enhancement of ventilation rate and thermal comfort.

Keyword: *Natural ventilation, classrooms retrofit, thermal comfort*

Criteria for partition placement to prevent ventilation deterioration

Miguel Yamamoto * [1], **Satoshi Hori**[2], **Shin-ichi Tanabe**[3], **Akihiro Kawamura**[1], **Hisashi Hasebe**[1], **Keichi Suzuki**[1]

[1] *Shimizu Corporation*

[2] *Juntendo University*

[3] *Waseda University*

During the COVID-19 pandemic, partitions were often placed on tables to prevent infection. These partitions can interfere with room ventilation, but the proper placement of partitions has not been clarified. This study aims to clarify the placement of partitions to avoid creating extreme ventilation deterioration. In this study, the age of air in a simplified space with partitions in various arrangements was analyzed by CFD (computational fluid dynamics). The validity of the CFD was confirmed by model experiments. The results showed that regarding the size of the gap between the partition and the ceiling, the smaller the gap, the larger age of air tended to be. Regarding wind direction, the more orthogonal the airflow between the partition and ventilation (the larger the wind direction), the larger age of air tended to be. Age of air increased significantly between 30° and 45° wind direction when the gap was more than 800 mm and between 15° and 30° wind direction when the gap was less than 600 mm. Based on these results, the following criteria were suggested for partition placement to avoid creating extreme ventilation deterioration: 1. The angle between the partition and wind direction should be less than 30°; 2. Especially, when the gap is less than 600 mm, the angle between the partition and wind direction should be less than 15°.

Keyword: *age of air, computational fluid dynamics, particle image velocimetry*

Advancing UV Disinfection: Insights and Optimization Strategies from CFD based UV dose model

Sudhanshu Pandey * , **Hannu Salmela, Aku Karvinen, Jani Hakala**
VTT Technical Research Centre of Finland Ltd., Espoo, Finland

This study develops a robust CFD model for UV disinfection systems, focusing on microbial inactivation and Reduction Equivalent Doses (RED). The computational domain represents a canonical room (7.25 m × 4.4 m × 2.8 m), with flow introduced through two ceiling-mounted diffusers and a 20 KW Far-UVC lamp centrally installed. A user-defined function calculates precise UV dosages, enabling accurate microbial inactivation predictions.

The model investigates key parameters, including UV transmittance (80–90%), lamp efficiency (90–100%), and inlet velocity (3-4 m/s), to evaluate their influence on disinfection efficacy. Validation against a previous study shows close agreement, with a maximum 2% difference in fluence rate distribution. Dose-response and frequency-dose curves are generated to analyze disinfection outcomes across various scenarios.

Preliminary results suggest that under conditions with 80% UV transmittance, 95% lamp efficiency, and inlet velocity of 3 m/s, RED reaches 1.5 mJ/cm², achieving microbial inactivation rates of 78–82% (1–1.1 log reduction). Moreover, with optimized parameters (90% transmittance, 100% lamp efficiency, and 4 m/s inlet velocity), the RED increases to 2.8 mJ/cm², yielding inactivation rates of 92–96% (1.5–1.7 log reduction). Scenarios with lamp efficiency reduced to 90% and UV transmittance at 80% show a decrease in RED to 1.2 mJ/cm², resulting in microbial inactivation rates of 65–70% (0.8–1 log reduction), demonstrating the need for higher system efficiencies to ensure effective disinfection.

The study emphasizes optimizing UV dose, exposure time, and system efficiency to maximize microbial inactivation while minimizing energy consumption.

Keyword: *Airborne microorganism, UV disinfection, CFD*

Evaluating the Influence of Ventilation Type and Fixture Placement on the Effectiveness of Upper Room UVGI in Reducing Airborne Infectious Aerosol Concentration in a Classroom

Youngbo Won[1], **William Bahnfleth** [*2]

[1] *California Department of Public Health*

[2] *The Pennsylvania State University*

Ultraviolet germicidal irradiation (UVGI) systems (also called Germicidal Ultraviolet, GUV) are commonly employed to reduce indoor airborne infection transmission. The susceptibility to germicidal wavelengths, primarily in the UVC band, varies from one pathogen to another. Factors such as temperature and humidity may affect their survival outside the host. This, in turn, impacts the distribution of active infectious aerosol and the exposure of occupants. This study investigated how the disinfection rates of two pathogens (SARS-CoV-2 and M. tuberculosis) exposed to 254 nm UVC radiation are influenced by typical indoor environmental factors. Using a computational fluid dynamics (CFD) model, we simulated a classroom with 12 students and one teacher to assess the concentration of infectious aerosols under various conditions, including different ventilation strategies (mixing and displacement) and UVGI fixture placement (ranging from 2 to 6 fixtures with different orientations). When two fixtures were used, UVGI effectiveness was enhanced by positioning the units along the longer walls of the classroom as opposed to the shorter front and back walls. UVGI combined with mixing ventilation was 35% more effective at inactivating SARS-CoV-2 than with displacement ventilation. In contrast, for tuberculosis, whose aerosol particles have a longer lifespan, and which has a higher resistance to UVGI, the system with mixing ventilation showed only a marginal advantage (6%) over displacement ventilation. This study highlights the significant impact of UVGI fixture configuration and ventilation strategy on viral disinfection effectiveness, particularly for short-lived, highly UV-susceptible pathogens like SARS-CoV-2.

Keyword: *CFD, Germicidal Ultraviolet, Ventilation*

Assessment of Factors affecting Indoor Microbiome Pollution in Schools

Akanksha Pharande, Sujata Karve *

Dr. Bhanuben Nanavati College of Architecture, Pune, Maharashtra, India

Air quality of indoor environments is one of the major aspects influencing health, wellbeing and productivity of people. Indoor air quality is affected by the presence of harmful microorganisms which include bacteria, fungi and viruses. This study aims to assess the extent of indoor air bacterial and fungal contamination present in classrooms of three selected schools, as well as to determine the relationship between microbial density, VOCs, PM 2.5, PM 10 and physical factors such as occupancy, temperature and relative humidity. Three schools were selected based on location, design, materials used and maintenance. The study was conducted in selected classrooms and corridors where air samples were collected with a microbial air sampler. Physical measurements of PM2.5, PM 10 and climatic parameters were also taken with the help of air quality meters and anemometer for wind. Sampling was done during school hours in the month of March with the samples taken at breathing zones. Correlation analysis was done to identify factors that contributed to the bacterial contamination.

The data was assessed against the sanitary standards of European Commission and level of contamination in school 2 was rated as intermediate. Compact planning with poor cross ventilation was found to be one of major causes of higher contamination.

The findings revealed that higher microbial pollution was affected by poor ventilation rates, layout of the spaces, materials used for interior spaces, higher humidity levels, outdoor contamination levels, occupancy and poor maintenance of the building.

Keyword: *Biological contaminants, microbiomes, indoor air quality*

Possible inter-room aerosol-mediated COVID-19 transmission in a quarantine hotel: a case study

Merrill Lim * [1], **Junjing Yang**[1], **Sze Tat Tan**[1], **Man Xiao**[1], **Chun Ping Gao**[2], **Vernon Lee**[3], **Yee Leong Teoh**[3], **James Chan**[3], **Lin Cui**[3], **Yichen Ding**[3], **Joanna Shen**[1], **Ruixin Li**[2], **Jang-Young Lee**[2], **Alvin Seoh**[2], **Shuzhen Sim**[1], **Lee Ching Ng**[1]

[1] *National environment agency Singapore*

[2] *Building and Construction Authority Singapore*

[3] *Ministry of Health of Singapore*

SARS-CoV-2 infections were reported in two individuals who stayed in separate rooms, one directly above the other, in a designated quarantine hotel in 2021. Epidemiological info and whole-genome sequencing analysis suggested that the two cases were linked, although they had no physical contact with each other.

We studied the possible route and cause of vertical inter-room transmission through a series of investigative tests, including dosing with aerosols and visual inspections. We also evaluated the potential of air purifiers as a mitigation measure by testing their effectiveness at removing aerosols in the recipient room.

Our investigation suggested the vertical inter-room transmission was due to the backflow of air from the exhaust system into the recipient room. Aerosols dosed in the source room travelled to the recipient room and the route was found to be likely via the toilet exhaust of the donor room into the exhaust shaft, which backflow into the plenum space above the ceiling of the room via the observed gaps and cracks in the brick wall of the exhaust shaft. Aided by the fan coil unit which draws air from the plenum space, the aerosols further travelled into the plenum space of bedroom compartment and eventually into the recipient room. The risk could have been further exacerbated by the poor exhaust air velocity in the recipient room. The results, taken together with whole-genome sequencing data and ruling out other potential epidemiological links, suggests inter-room COVID-19 aerosol transmission. Use of an air purifier equipped with a high-efficiency particulate air (HEPA) filter in the recipient room accelerated aerosol decay, providing evidence for their use as a mitigation measure to reduce risk of exposure to virus aerosols in an indoor environment.

Keyword: *SARS-Cov-2, Aerosol, Infectious disease*

Modulation of Indoor Microbiota through the Introduction of Indoor Plants

<u>Supreet Kaur</u> * [1], Dr. Sumit Kumar Mishra[2], Dr. Venkata Ramana[3], Madhu Tiwari[3], Kartika Pandey[1], Mamta Devi[1], Dr. Randeep Guleria[4]

[1] *CSIR National Physical Laboratory, Dr. K.S. Krishnan Marg, New Delhi-110012, India, Academy of Scientific and Innovative Research (AcSIR), Ghaziabad - 201002, Uttar Pradesh, India*

[2] *CSIR National Physical Laboratory, Dr. K.S. Krishnan Marg, New Delhi-110012, India, Academy of Scientific and Innovative Research (AcSIR), Ghaziabad - 201002, Uttar Pradesh, India, Department of Chemistry, University of British Columbia, V6T 1Z1, Canada*

[3] *Microbiological Type Culture Collection and Gene Bank (MTCC), CSIR Institute of Microbial Technology, Academy of Scientific and Innovative Research (AcSIR), Ghaziabad - 201002, Uttar Pradesh, India*

[4] *Institute of Internal Medicine & Respiratory and Sleep Medicine, India, All India Institute of Medical Sciences, New Delhi*

Indoor plants are increasingly incorporated into built environments due to their potential to improve indoor air quality (IAQ), reduce particulate matter (PM) concentrations (Kaur et al., 2025), and enhance psychological well-being. While their role in air purification has been widely examined, there remains a critical gap in understanding their influence on indoor microbial communities. This study investigates how the presence of indoor plants alters the composition and abundance of microbial communities in indoor air and on surfaces. A controlled six-month experiment was conducted at the CSIR–National Physical Laboratory (28.70°N, 77.10°E) in Delhi, India, where selected indoor plant species were introduced into a vegetated room and compared with a comparable reference room without vegetation.

The vegetated room exhibited a 4.4% reduction in temperature and a 25.1% increase in relative humidity, indicating plant-induced microclimatic changes. Microbial community analysis revealed a shift in composition, with increased representation of potentially beneficial genera such as Bacillus, Paenibacillus, and Lysinibacillus. Jaccard similarity indices—used to quantify community overlap—indicated low similarity between plant-associated microbes and those in air (0.30) and on surfaces (0.36), while a modest overlap between soil and indoor surfaces (0.57) suggests limited microbial transfer from soil reservoirs.

These findings suggest that while indoor vegetation may enhance microbial diversity and introduce beneficial taxa, the extent of microbial dispersion is minimal and primarily soil-mediated. Thus, to optimize indoor ecological quality while maintaining microbial safety, future greening strategies should incorporate sterile or managed soil systems.

References:

Kaur, S., Mishra, S. K., Goel, V., Kumar, M., Singh, R., Devi, M., ... & Kumar, P. (2025). Impact of indoor plant-induced relative humidity on PM concentration in indoor urban environment. Atmospheric pollution research, 16(6), 102468.

Keyword: *Indoor environment, Bacteria, Indoor Plants*

Assessing the placement of HEPA air cleaners in a multi-zone hospital ward to reduce infection risk

Anushi Khandare * [1], **Alexander Edwards**[2], **Catherine Naokes**[1], **Martin Lopez-Garcia**[1], **Marco-Felipe King**[1], **Louise Fletcher**[1], **Andrew Carnegie**[3]
[1] *University of Leeds, Leeds, United Kingdom LS2 9JT*
[2] *University of Bristol, Bristol, United Kingdom BS8 1QU*
[3] *Air Sentry Limited Unit 1, Wiltshire, SN15 5EH*

The COVID-19 pandemic has emphasised the need for understanding the dynamics of airborne infections and good ventilation, particularly in a hospital. When deployed safely and maintained well, air cleaners are an important advancement in mitigating against indoor air pollution and infection transmission, especially where increasing ventilation is challenging. This work investigates the impact of air cleaners on reducing the concentration of airborne pathogens in a multi-zone naturally ventilated hospital ward using CONTAM modelling. The methodology and the ward design were based on prior work on ventilation by Edwards et.al [2023]. The model assumed HEPA-based air cleaners were strategically placed at different locations in the hospital ward zones with an infectious person moving between spaces. Simulations analysed fluctuations in pathogen concentration and occupant exposure over an 8-hour period. The reduction in concentration was strongly influenced by the location of air cleaners relative to the pathogen source and the ventilation pattern along with spatial arrangement of each zone. Results demonstrate air cleaners can significantly reduce airborne pathogen concentrations, but where the location of an infectious person is changing or uncertain, more devices may be needed. The model provides useful insights for designing multi-zone wards to minimize airborne transmission and improve safety for healthcare workers and patients. CONTAM is effective for exploring zonal flows and time fluctuations. However, as the model focuses on bulk flow, validating the results with experimental data, and detailed analysis of airflow patterns using computational fluid dynamics to explore spatial variations will form the basis of future work.

References:

Edwards, A.J., Benson, L., Guo, Z., Lopez-García, ´ M., Noakes, C.J., Peckham, D., et al., 2023. A mathematical model for assessing transient airborne infection risks in a multi-zone hospital ward. Build Environ 238, 110344

Keyword: *Airborne infections, air cleaners, CONTAM*

Development and Performance Evaluation of a Ductless Air Handling Unit for Airborne Infection Control in Critical Care Environments

Zakir Hussain *
Founder

This study presents the design, development, and performance evaluation of a patented ductless air handling unit (AHU) system for conditioning and sterilizing air in critical hospital environments. The system integrates HEPA and ULPA filtration, UV-C sterilization, ozone sterilization, cooling, heating, and laminar airflow without ducting, ensuring Class I air quality standards (>10-100 particles per cu. ft.) in operation theatres (OTs), intensive care units (ICUs), and neonatal intensive care units (NICUs). The system demonstrates significant advantages over conventional ducted AHU systems in terms of maintenance, energy savings, cost-effectiveness, and infection control efficiency. The system has been successfully installed in facilities such as AIIMS Raebareli, demonstrating real-world applicability.

Keyword: *Ductless AHU, Airborne Infection Control, Hospital Ventilation*

IMPACT OF ADVANCED FILTER MEDIA ON INDOOR AIR QUALITY AND ENERGY PERFORMANCE OF BUILDINGS IN REGIONS WITH HIGH PARTICULATE MATTER POLLUTION

Pooja Shukla *, **Shashank Mishra, Mohammed Umar**
GBCI India, Noida, India

In high particulate matter (PM) regions, installing high-efficiency filters in the building HVAC system to filter PM from incoming air is crucial for occupant health. Due to higher resistance and pressure drop than lower efficiency filters, high-efficiency filters impact energy consumption. However, advanced filter media technology offers high-efficiency, low-pressure drop alternatives.

This field research retrofitted two AHUs in an office building in Noida, India, with two types of advanced media filters, and compared their performance with the building's existing filters. All three filters—existing and two new advanced—had MERV 14 rating but were manufactured using media with different specifications. Performance of the filters in terms of their PM filtration efficacy, particularly during high outdoor PM pollution; and impact on energy consumption, was evaluated.

Field data for air quality and energy consumption were collected for around five months post-retrofit (January to May 2024) through spot testing and continuous monitoring for evaluating filters' performance. Energy modelling was conducted to determine filters' impact on fan energy consumption.

All three filters demonstrated high filtration efficiency, but the advanced electret media filter (with single charged layer, triboelectric effects between dissimilar fibres) performed best in terms of PM filtration and energy impact. It maintained mean indoor PM2.5 level of 2.4 $\mu g/m^3$ with mean outdoor level of 85 $\mu g/m^3$ and could reduce fan energy consumption by 29% compared to existing filters.

The findings highlight that filter media characteristics impact filter performance and that technologically advanced filters can deliver healthy IAQ in high PM regions, while optimizing energy performance.

Keyword: *particulate matter (PM), indoor air quality (IAQ), advanced filter media technology*

Navigating Indoor Air Pollution: Research Insights on Understanding Pollutants, Mitigation Strategies, and Emerging Technologies.

Janmejoy Gupta, Neha MP *
School of Planning and Architecture, Vijayawada

Indoor air quality (IAQ) has become a critical determinant of human health and productivity, given that individuals spend a significant portion of their lives indoors. This exposure heightens the risk of inhaling a variety of pollutants, originating from human activities, outdoor contamination, and the use of materials and products within indoor environments. These pollutants like particulate matter, volatile organic compounds etc, can infiltrate deep into the respiratory system, impair cognitive functions and adversely affect overall well-being. Alarmingly, indoor pollutant levels can often surpass those found outdoors further emphasizing the need for effective mitigation strategies.

This research explores the intricate relationships between building characteristics, occupant activities, and the accumulation of indoor pollutants. By reviewing scientific studies conducted globally, we identify prevailing trends, research gaps and emerging challenges in the field of IAQ. The study emphasizes the necessity of addressing these challenges to create healthier indoor spaces.

The findings highlight the importance of evidence-based strategies to improve IAQ, including the adoption of advanced ventilation systems, low-emission building materials and innovative air cleaning technologies. This research aims to formulate evidence-based strategies for improving indoor environments. By addressing the multifaceted causes of poor IAQ and implementing these targeted solutions, it is possible to create healthier, more sustainable indoor environments that significantly enhance the quality of life for occupants. This research underscores the necessity of integrating these measures into building design and operation to foster long-term well-being and productivity.

Keyword: *Indoor air quality, sustainable indoor environment, low-emission*

Navigating Indoor Air Pollution: Research Insights on Understanding Pollutants, Mitigation Strategies, and Emerging Technologies.

Neha MP, Dr. Janmejoy Gupta

School of Planning and Architecture – Vijayawada, India

nehamp1205@gmail.com.

SUMMARY:

Indoor air quality (IAQ) is vital for human health and well-being, as people spend a large amount of time indoors. Indoor pollutants, including particulate matter and volatile organic compounds, can negatively impact respiratory health, cognitive function, and overall well-being. Indoor pollutant levels often exceed outdoor levels, highlighting the need for mitigation.

This research examines the relationship between building factors, occupant activities, and indoor pollutant accumulation. By reviewing global studies, it identifies trends and challenges in IAQ research. Addressing these challenges is essential for creating healthier indoor spaces.

The findings emphasize evidence-based strategies to improve IAQ, such as advanced ventilation, low-emission materials, and air-cleaning technologies. This research aims to develop such strategies to enhance indoor environments. By addressing the causes of poor IAQ and implementing these solutions, healthier, more sustainable indoor spaces can be created, improving occupants' quality of life. Integrating these measures into building design and operation is necessary for long-term well-being and productivity.

In essence, this emphasizes that good indoor air is vital and that we need research-backed ways to achieve it in buildings.

Keywords: Indoor plants; Indoor Air Quality; hostel rooms; occupant well-being, sustainable indoor spaces, Sick Building Syndrome, aerosols, particulate matter,

INTRODUCTION:

The vital importance of indoor air quality

The quality of indoor air has become a pressing concern in recent years. With people globally spending a substantial portion of their lives indoors, the nature of the indoor environment, particularly the air that is breathed, plays a pivotal role in overall health and well-being. The air we breathe indoors plays a crucial role in our health and well-being, given the significant amount of time we spend in indoor spaces globally. Poor quality of indoor air has been associated with a range of health issues, from immediate discomfort to the development of Sick Building Syndrome (SBS) and the aggravation of chronic respiratory and cardiovascular conditions. It is essential to recognize that the way buildings are operated and the materials used in their construction not only impact indoor air quality but also have far-reaching environmental consequences that contribute to broader pollution challenges beyond the confines of indoor spaces. Poor indoor air quality (IAQ) is directly linked to the development of Sick Building Syndrome (SBS) and the exacerbation of long-term respiratory and cardiovascular diseases. Furthermore, the operational aspects of buildings and the materials chosen for their construction have considerable environmental impacts, contributing to broader pollution issues that extend beyond the immediate indoor environment. This paper aims to provide a comprehensive understanding of the common pollutants found in indoor air, their detrimental effects, and the various strategies and technological solutions available for their mitigation. By exploring current research, this report will delve into the principles of building design that promote healthy air, the efficacy of existing air cleaning technologies, and the promise of emerging innovations in monitoring and purification. Ultimately, the goal is to synthesize this information into a cohesive overview that empowers building owners, facility managers, architects, and health professionals to create healthier indoor spaces, thereby preventing SBS and minimizing the environmental footprint of the built environment.

Understanding the Spectrum of Indoor Air Pollutants

The air within buildings can contain a complex mixture of pollutants, each with its own sources, characteristics, and potential health effects. Understanding these pollutants is the first step towards effectively mitigating their impact. The primary categories of concern include aerosols, particulate matter (specifically PM2.5 and PM10), and volatile organic compounds (VOCs).

Aerosols: Definition, Sources, and Behaviour in Indoor Environments

Indoors, aerosols are microscopic solid or liquid particles suspended in air, often synonymous with fine particulate matter. They originate from combustion (oil, gas, wood, tobacco), building materials (asbestos, damp carpets, pressed wood), household products (cleaners, air fresheners), human activities (cooking, smoking, hobbies), infiltrated outdoor air (pollen, dust, exhaust, smoke), biological contaminants (dander, mold, dust mites), and office equipment. Aerosol behaviour is dynamic; particle distribution changes and sinking depends on the size and density. They can carry other pollutants, depositing them in the respiratory system and on surfaces.

Particulate matter (pm2.5 and pm10): Sources, Characteristics, and Health Implications for Building Occupants

Particulate matter (PM) includes inhalable coarse particles (PM10, $\leq$ 10 μm) and fine particles (PM2.5, $\leq$ 2.5 μm, within PM10). Indoor sources include infiltrated outdoor pollution (vehicle exhaust, industrial emissions, wildfires), indoor combustion (cooking, candles, fireplaces, heaters), tobacco smoking, tracked-in dust, biological contaminants, cleaning, hobbies (woodworking, arts), and office equipment. PM's complex composition includes nitrates, sulphates, organics, metals, and allergens. PM2.5 can deeply penetrate lungs and the bloodstream, causing respiratory and cardiovascular issues. PM10 irritates eyes, nose, and throat. Both can worsen asthma, increase infection susceptibility, contribute to cardiovascular events, and long-term exposure is linked to reduced lung function and chronic diseases.

Volatile organic compounds (vocs): Common Sources, Chemical Diversity, and Impact on Health

Volatile organic compounds (VOCs) are diverse organic chemicals emitted as gases from solids or liquids at room temperature, often at higher indoor concentrations than outdoors. Common indoor sources include building materials (paints, adhesives, flooring, pressed wood), household/personal care products (cleaners, air fresheners, cosmetics), activities (smoking, dry cleaning, crafts, cooking, fuel-burning stoves, office equipment), and furnishings (upholstery, treated fabrics). VOCs have extensive chemical diversity (e.g., benzene, formaldehyde). Short-term exposure can cause irritation, headaches, nausea, and breathing difficulties. On exposure for long-term it, may increase the risk of causing liver, kidney, and central nervous system damage; some VOCs are carcinogens and can worsen respiratory conditions.

The tangible consequences:

Sick Building Syndrome and Health Effects

Exposure to the spectrum of indoor air pollutants can lead to a variety of adverse health outcomes, including the phenomenon known as Sick Building Syndrome (SBS) and other broader health effects.

Defining Sick Building Syndrome (SBS)

Sick Building Syndrome (SBS) occurs when occupants experience acute health and comfort problems linked to the amount of time spent in a building. Unlike Building Related Illness (BRI), SBS lacks a single, identifiable cause or illness. Symptoms, which can be localized or widespread, like headache, eye/nose/throat irritation, dry cough, dry/itchy skin, dizziness/nausea, concentration difficulties, fatigue, and odour sensitivity. Other reported symptoms are muscle/joint pain, congestion, and skin rashes. Diagnosing SBS is difficult due to its non-specific, overlapping symptoms and the absence of a definitive test, requiring the ruling out of other causes and assessment of the indoor environment.

SBS often results from the complex interaction of indoor environmental factors, with indoor air pollutants being significant contributors. Poor ventilation is a major factor, facilitating the buildup of pollutants. Chemical contaminants like VOCs from indoor sources (building materials, office equipment, pesticides, cleaners) and environmental tobacco smoke are strongly implicated. Outdoor chemical contaminants (vehicle exhaust, etc.) entering through poor ventilation can also contribute. Biological contaminants (bacteria, molds, pollen, viruses) in poorly maintained, moist environments can trigger SBS-like symptoms. Beyond air quality, physical factors such as noise, vibrations, electromagnetic fields, poor temperature control, insufficient natural light, and artificial lighting can also contribute to SBS.

Prolonged exposure to poor indoor air quality can also lead to more severe and long-lasting health consequences beyond the acute symptoms of SBS. Poor IAQ can directly and significantly contribute to infections, lung cancer, and chronic lung diseases. Individuals with pre-existing lung conditions are at higher risk for exacerbated symptoms and adverse effects. Exposure to various indoor air pollutants can cause respiratory problems, trigger or worsen allergies, and aggravate cardiovascular conditions. Long-term exposure to specific pollutants like radon and asbestos can cause serious diseases, including lung cancer and other respiratory illnesses, which may manifest years later. The vulnerable people, including children, the elderly, and those with pre-existing respiratory or cardiovascular ailments, are particularly susceptible to the health effects of poor IAQ. Furthermore, poor indoor air quality can impact cognitive function, leading to concentration difficulties and reduced productivity, and affect overall well-being.

Methodology: The study employed a field experiment conducted over eight weeks to assess the impact of indoor plants on indoor air quality in hostel rooms. Three rooms of equal size, orientation, and opening dimensions were selected, each assigned a different plant species: Spider Plant, Snake Plant, and Areca Palm. Air quality parameters, including CO2, HCHO, TVOC, temperature, and relative humidity, were measured weekly using an air quality monitor. The number of plants in each room was gradually increased over the weeks to observe the effect of varying foliage density on air quality.

The study focused on measuring several specific indoor air pollutants to assess the effectiveness of indoor plants in enhancing air quality within hostel rooms. The pollutants measured included:

- **Carbon Dioxide (CO_2)**: A common indoor pollutant that can accumulate due to human activity and poor ventilation.
- **Formaldehyde (HCHO)**: A volatile organic compound (VOC) often found in building materials and household products, known for its potential health risks.
- **Total Volatile Organic Compounds (TVOC)**: A group of organic chemicals that can have short- and long-term adverse health effects.
- **Temperature**: While not a pollutant, temperature is an important environmental factor that can influence indoor air quality and comfort.
- **Relative Humidity**: Similar to temperature, humidity levels can affect air quality and the performance of indoor plants in purifying the air.

These measurements were taken using an indoor air quality monitor to evaluate the impact of different plant species on the indoor air quality and overall well-being of the occupants in the hostel rooms.

Pilot Survey

The School of Planning and Architecture (SPA) Vijayawada's South Block Girls' Hostel accommodates approximately 327 students. The hostel features shared rooms and single rooms. pilot survey is done to select single rooms for the study.

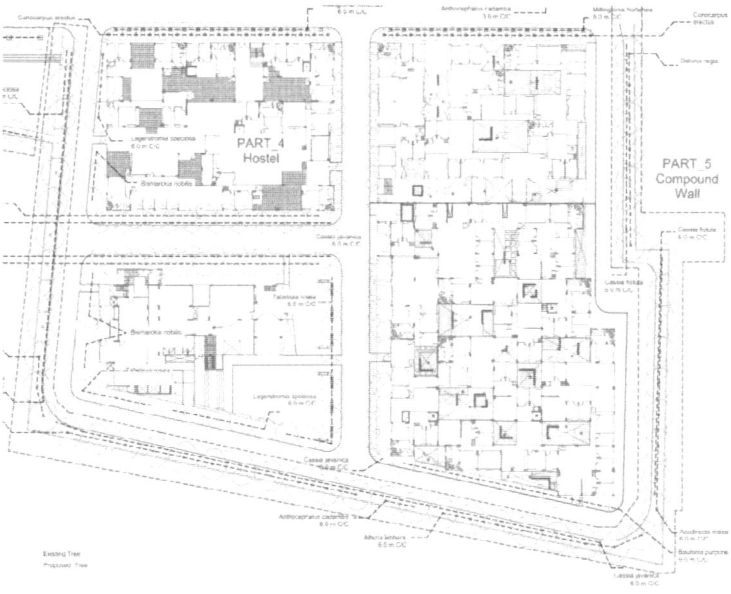

Figure 1: Plan of Accommodation – hostels in SPA-V campus.

PILOT SURVEY									
ROOMS	NAME	Air quality index	POLLUTANTS				AQ Indicat	env parameter	
		AQI	**HCHO**	**TVOC**	**PM2.5**	**CO**	**CO2**	**TEMP**	**HUMIDITY**
ROOM 1	Aumgana	15	0.007	0	21		401	29	64
ROOM 2	Deepika	21	0.006	0	30		400	30	55%
ROOM 3	Divya	12	0.004	0.029	17	1	536	28	61
ROOM 4	Sharon	7	0.012	0	11		402	29	62
ROOM 5	Sharvi	16	0.003	0.05	23		400	30	55
ROOM 6	Sweta	21	0.007	0	30		400	30	57%
ROOM 7	Shanthini	16	0.003	0.229	23	2	810	31	55%
ROOM 8	Sakshi	7	0.019	0.006	10		400	28	69
ROOM 9	Ponmani	19	0.032	0.045	27	1	575	28	57
ROOM 10	kreti	15	0.008	0	21		402	29	64
ROOM 11	Isha	19	0.003	0.444	27	2	896	26	70
ROOM 12	Liyana	7	0.019	0.006	11		402	28	69
ROOM 13	malavika	7	0.019	0.038	11	1	555	28	61
ROOM 14	rishika	9	0.022	0.038	13	1	596	28	62
ROOM 15	Vrindha	10	0.005	0.032	14		405	29	61
ROOM 16	Shreya	7	0.019	0.006	10		402	28	69
ROOM 17	Jubi	7	0.019	0.04	11		410	28	69
ROOM 18	thanha	7	0.019	0.006	10		404	28	64
ROOM 19	zainaba	7	0.012	0.031	11		402	29	62
ROOM 20	ann	14	0.012	0	23		402	29	64

Table1: Pilot survey details

RESULTS:

Air Quality Concerns: PM2.5 levels exceed safe limits in Room 1, Room 3, Room 5, Room 7, Room 9,

• Room 11, and Room 20, indicating poor air quality.

• Ventilation Status: CO_2 levels are within acceptable limits in all rooms, ensuring adequate ventilation.

• Chemical Pollutants: Formaldehyde levels exceed the WELL standard in Room 9, requiring attention.

Effectiveness of Indoor Plants:

The study found that certain indoor plants, particularly snake plants, were effective in improving IAQ by reducing specific pollutants such as CO_2 and formaldehyde. This aligns with previous research indicating that indoor plants can significantly lower CO_2 concentrations in controlled settings by up to 20%.

Pollutant Reduction: The measurements indicated that the presence of indoor plants contributed to a noticeable reduction in the levels of CO_2, HCHO, and TVOC in the hostel rooms. The exact percentage reductions were not specified in the provided contexts, but the implication is that the plants played a crucial role in purifying the air

Variability Based on Plant Species: The effectiveness of pollutant removal varied based on the species of plants used. Different plants have different capacities for absorbing pollutants, which suggests that selecting the right species is essential for maximizing air quality benefits.

Environmental Conditions: The study highlighted those environmental conditions, such as temperature and humidity, also influenced the performance of the plants in improving IAQ.

The warm and humid climate of Vijayawada provided an ideal setting for studying these effects, as it is conducive to plant growth and air purification.

Occupant Well-being: Beyond pollutant reduction, the presence of indoor plants was associated with improved occupant well-being. This includes psychological benefits, such as reduced stress and enhanced comfort levels, which are important in confined living spaces like hostel rooms

Cost-Effectiveness: Integrating indoor plants into hostel rooms was identified as a cost-effective strategy to enhance environmental quality. This suggests that even a small number of plants can have a massive positive impact on the air quality and occupant health, making it a practical solution for improving living conditions in hostels

DISCUSSIONS:

Beyond occupant health:

Environmental Ramifications of Indoor Air Pollution

The consequences of poor indoor air quality goes beyond just the health of building occupants, reaching into broader environmental issues and impacting the sustainability of the built environment.

Indoor Pollutants' Contribution to Outdoor Air Pollution

Indoor air pollution significantly impacts outdoor air quality and regional environments. Burning fuels indoors releases pollutants like particulate matter and VOCs that escape outside. Household air pollution, combined with other sources, contributes to many premature deaths globally. VOCs released indoors can react in the atmosphere, forming ground-level ozone and smog, highlighting the need for indoor source control and filtration.

Life Cycle Impacts of Building Materials and HVAC on Sustainability

The environmental impact of indoor air pollution extends to the life cycle of building materials and HVAC systems. Building materials can release pollutants throughout their lifecycle. Energy-intensive HVAC systems contribute to energy demand and greenhouse gas emissions. Refrigerants can also harm the environment. Sustainable building prioritizes low-VOC materials and energy-efficient HVAC to minimize environmental impact and improve indoor air quality.

Designing for healthy air:

Proactive Building Design Strategies

Creating buildings that promote good indoor air quality requires a proactive approach that integrates air quality considerations into the fundamental design and construction processes. This involves optimizing ventilation systems, strategically selecting building materials, leveraging natural airflow, considering building layout and zoning, and controlling humidity.

Optimizing Ventilation Systems: Natural, Mechanical, and Hybrid Approaches for Effective Air Exchange

Adequate ventilation is a cornerstone of good indoor air quality, playing a critical role in diluting and removing the various pollutants that can accumulate within buildings. Ventilation can be achieved through natural means, mechanical systems, or a combination of both in hybrid approaches. Natural ventilation relies on the movement of air through openings in the building envelope, such as windows and doors. When outdoor air quality is acceptable, opening windows and doors can effectively increase airflow and help to flush out indoor pollutants. Effective design for natural ventilation involves careful consideration and design decisions of the building's orientation relative to prevailing winds and the strategic sizing and placement of these openings to maximize cross-ventilation and airflow.

Mechanical ventilation systems provide a more controlled means of air exchange, utilizing fans to either exhaust stale indoor air, supply fresh outdoor air, or both in balanced systems. These systems can range from simple exhaust fans in bathrooms and kitchens to more sophisticated whole-house systems that continuously supply filtered and conditioned outdoor air while exhausting contaminated indoor air. Balanced mechanical ventilation systems, particularly those incorporating heat recovery ventilators (HRVs) or energy recovery ventilators (ERVs), offer an energy-efficient solution by transferring heat or both heat and moisture between the incoming and outgoing air streams, reducing the energy load associated with conditioning the ventilation air. ASHRAE Standard 62.2 specifies specific guidance on ventilation and acceptable indoor air quality in low-rise residential buildings, serving as a nationally recognized standard for this purpose. Hybrid ventilation systems strategically combine elements of both natural and mechanical ventilation to optimize air quality and energy efficiency. Furthermore, the implementation of demand-controlled ventilation (DCV) systems, which utilize sensors to monitor occupancy levels and adjust ventilation rates accordingly, can help to reduce energy waste by adapting to the actual needs of the building occupants in real-time.

Strategic Material Selection:

Choosing low/zero-VOC building materials (paints, adhesives, flooring) is crucial for good indoor air quality. Prioritize options like solid wood, low-VOC plywood, and natural insulation. Seek low-emission certifications (GREENGUARD, LEED). Avoid formaldehyde and select moisture-resistant materials to prevent mold.

Passive Ventilation: Maximize natural airflow using building orientation and cross-ventilation (operable windows/vents). Utilize stack effect ventilation and architectural features like clerestories. Design and occupant participation are key.

Building Layout and Occupancy: Isolate pollutant sources (kitchens) with exhaust ventilation. Locate air intakes away from pollution. Ensure proper space planning and manage occupant density. Implement smoking bans.

Humidity Control: Maintain 30-60% humidity to prevent mold and dust mites. Use dehumidifiers and ensure proper drainage.

Air Cleaning Technologies: Air cleaning technologies include filtration systems, air purifiers, advanced ventilation, UVGI, ionizers, and electrostatic precipitators.

Air Filtration Systems: MERV ratings indicate filter effectiveness (higher is better). HEPA filters are highly efficient. Choose filters based on pollutants and replace them regularly.

Other existing technologies:

UVGI, Ionizers, Electrostatic Precipitators – Mechanisms and Evidence of Effectiveness:

In addition to filtration, other technologies are used for indoor air cleaning, including Ultraviolet Germicidal Irradiation (UVGI), ionizers, and electrostatic precipitators. UVGI technology utilizes ultraviolet (UV) light to inactivate microorganisms such as viruses, bacteria, and mold that are present in the air or growing on surfaces within the HVAC systems. The effectiveness of UVGI depends on the intensity of the UV light and the duration of exposure. Some UVGI devices may produce ozone as a byproduct, which is a potential health concern. Ionizers work by dispersing the electrically charged ions into air, which then attaches to the airborne particles, which charges them causing to clump together and settle out of the air or adhere to nearby surfaces. While ionizers can reduce airborne particles, they may also generate ozone and can lead to the deposition of particles on room surfaces, requiring more frequent cleaning. Electrostatic precipitators (ESPs) also use the electrical charge to remove particles present in the air. They draw air through an ionization where particles are charged and then collected on oppositely charged plates within the system. The efficiency of ESPs can decrease over time as the collector plates become loaded with particles, and some ESPs may also produce ozone. Generally, the Environmental Protection Agency (EPA) suggests that for particle removal in homes, filtration methods, particularly those using HEPA filters, are often a safer and more effective choice compared to ozone-generating devices like some UVGI units and ionizers.

Portable and Whole-House Air Purifiers: Portable air purifiers clean air in specific areas using filters and fans. Effectiveness depends on filter type (HEPA for particles, activated carbon for gases), Clean Air Delivery Rate (CADR), and usage. Whole-house purifiers integrate into HVAC systems for larger spaces. Consider room size, pollutants, noise, and filter costs when selecting a purifier.

Advanced Ventilation Systems: Modern HVAC systems can include advanced filtration (MERV 13+ filters) and purification. Energy/heat recovery ventilators (ERVs/HRVs) provide fresh air with filtration. Smart ventilation systems adjust ventilation and purification based on real-time air quality, optimizing air quality and energy use.

The Future of Clean Air: Emerging technologies are improving indoor air quality monitoring and pollution mitigation like advanced Sensors and AI Monitoring: Advanced, portable sensors with AI/ML accurately measure pollutants (PM, VOCs, CO2, ozone) in real-time. AI analyses data to predict issues and optimize HVAC systems, enabling automated air quality control.

Novel Filtration Materials: Advanced HEPA filters with activated carbon and nanoparticles trap particles and neutralize bacteria, viruses, and VOCs. Nanotechnology enhances electrostatic precipitators. New filters use ceramic technology and electro spun nanofibers. Sustainable alternatives like botanical and microalgae-based purifiers are also being developed.

Advanced Air Purification Methods: Photocatalytic oxidation (PCO) uses UV light and a catalyst to break down pollutants. Cold plasma and dry hydrogen peroxide (DHP) systems are also emerging. Smog-eating materials like photocatalytic coatings on surfaces can reduce outdoor pollution.

Actionable steps:

The Importance of Regular Maintenance, Cleaning, and Occupant Behavior in Maintaining Good IAQ

Maintaining good indoor air quality in existing buildings and homes is an ongoing process that requires regular maintenance, thorough cleaning, and the adoption of healthy occupant behaviours. Regular cleaning, including dusting with a microfiber or damp cloth and vacuuming with a vacuum cleaner that is equipped with a HEPA filter, is very essential for removing dust, allergens, and other particulate matter that can accumulate on surfaces and in carpets. Spills should be cleaned up promptly to prevent the growth of mold and bacteria. Maintaining optimal indoor humidity levels through the use of dehumidifiers in damp areas and ensuring adequate ventilation can also help to prevent mold growth and reduce dust mite populations. Proper storage and handling of food and waste are important for preventing pests and minimizing Odors. Educating building occupants about the various sources of indoor air pollution and the best practices for reducing their exposure is crucial for fostering a collective effort towards better IAQ. Regular inspection and maintenance of HVAC systems, which also includes the changing of filters and cleaning ducts, as well as maintaining portable air purifiers according to the manufacturer's instructions, are essential for ensuring that these systems function efficiently and contribute to improved air quality.

Enhancing Indoor Air Quality in Existing Buildings: Improving indoor air quality in existing buildings will need a multifaceted strategy encompassing source control, ventilation enhancement, strategic air purifier use, and diligent housekeeping.

Source Control as a Primary Strategy: Prioritizing source control implies using low/zero-VOC products, appropriate chemical storage, and also avoiding indoor combustion. Selecting formaldehyde-free materials and controlling biological pollutants through cleaning and humidity management is vital. Safe operation and venting of fuel-burning appliances are essential strategies as well.

Ventilation Enhancement for Pollutant Reduction: Enhancing ventilation, through natural means (opening windows) and mechanical methods (exhaust fans), dilutes pollutants. Optimizing airflow and considering ERVs improves air exchange.

Strategic Use of Air Purification

HEPA filter air purifiers, with appropriate CADR and certification, provide supplemental pollutant removal. Correct placement and filter maintenance are vital. Activated carbon filters address VOCs, while ozone-generating purifiers should be avoided.

CONCLUSION:

Towards a future of healthier and more sustainable indoor environments:

Indoor air pollution represents a substantial challenge to both human health and environmental sustainability. The air within our buildings, where we spend the majority of our lives, can be laden with a complex mixture of pollutants, including aerosols, particulate matter, and volatile organic compounds, each with the potential to cause or exacerbate a range of health issues, from the discomfort of sick building syndrome to the severity of chronic respiratory diseases and cancer. Furthermore, the sources and impacts of indoor air pollution extend beyond the confines of individual buildings, contributing to broader environmental problems through outdoor emissions and the life cycle of building materials and HVAC systems.

Addressing this multifaceted issue requires a comprehensive and proactive approach. As this report has outlined, effective mitigation strategies encompass a range of actions, starting with the fundamental principles of building design that prioritize ventilation, material selection, natural airflow, layout, and humidity control. Current air cleaning technologies, including advanced filtration systems and portable air purifiers, offer valuable tools for removing existing pollutants from indoor environments. The future holds even greater promise with the emergence of innovative solutions such as AI-powered air quality monitoring and novel air purification methods like photocatalytic oxidation, which are poised to revolutionize how we understand and manage indoor air quality.

For existing buildings and homes, practical strategies focusing on source control and elimination, enhancing ventilation through simple and accessible means, and the informed selection and utilization of air purifiers and HVAC filters can yield significant improvements in indoor air quality. Moreover, the ongoing importance of regular maintenance, thorough cleaning, and the cultivation of healthy occupant behaviours cannot be overstated in sustaining these improvements over time.

Ultimately, navigating the challenges of indoor air pollution and striving towards a future of healthier and more sustainable indoor environments demands a multi-faceted effort. This includes continued research and innovation in building science and air purification technologies, the widespread adoption of best practices in building design and maintenance, the implementation of supportive policies and regulations, and an increased awareness and engagement from building owners, facility managers, and occupants alike. By embracing a proactive and holistic approach, we can collectively work towards creating indoor spaces that not only protect our health but also help in developing a more sustainable and environmentally responsible future.

ACKNOWLEDGEMENT:

I would like to express my sincere gratitude to Dr. Janmejoy Gupta, Associate Professor at the School of Planning and Architecture, Vijayawada, for his invaluable guidance, unwavering encouragement, and dedicated support throughout the research and writing process. His pursuit of excellence and constant support were instrumental in the thorough research and successful completion of this paper.

I also extend my appreciation to the School of Planning and Architecture, Vijayawada, its director, and the Programme Coordinator of the Department of Sustainable Architecture for providing this valuable opportunity to the students to engage in such impactful research.

Special thanks are due to Liyana Fathima for her dedicated work in researching the assessment of air quality in indoor spaces and the efficacy of strategies, including the use of indoor plants, to improve it. Her diligent efforts and insightful analysis significantly contributed to the findings and results presented in this paper.

Finally, I wish to acknowledge the unwavering support of my parents and friends, Rishika Shrivastava and Ajay. Their help and constant encouragement were invaluable throughout this endeavour.

REFERENCES:

1. **Delgado-Saborit, J. M.** (2024). Pollutants from Gas Stoves Kill 40,000 Europeans Each Year, Report Finds. *The Guardian*. https://www.theguardian.com/environment/2024/oct/28/pollutants-from-gas-stoves-kill-40000-europeans-each-year-report-finds
2. **Eftekhari, A., et al.** (2023). Chemistry of Indoor Air Pollution. *American Chemical Society*. https://pubs.acs.org/doi/book/10.1021/acsinfocus.7e7018
3. **Indoor Air Quality Society**. (2023). Indoor Exposure to Selected Air Pollutants in the Home Environment: A Systematic Review. https://iaqsociety.org/2023/05/05/indoor-exposure-to-selected-air-pollutants-in-the-home-environment-a-systematic-review.
4. **Mannan, M., & Al-Ghamdi, S. G.** (2021). Indoor Air Quality in Buildings: A Comprehensive Review on the Factors Influencing Air Pollution in Residential and Commercial Structures. *International Journal of Environmental Research and Public Health*, 18(6), 3276. https://doi.org/10.3390/ijerph18063276.
5. **Mishra, A. K., & Coggins, M.** (2024). Festive Scented Candles Could Be Worse for Your Health Than Cigarettes. *The Sun*. https://www.thesun.ie/health/14332487/festive-scented-candles-health-cigarettes-christmas/
6. **Osuizugbo, I. C., et al.** (2025). Emerging Technologies for Mitigating Air Pollution in Buildings: Systematic Review and Meta-Analysis. *Built Environment Project and Asset Management*. https://doi.org/10.1108/BEPAM-03-2024-0050
7. **Spengler, J., & Adamkiewicz, G.** (2009). Indoor Air Pollution: An Old Problem with New Challenges. *International Journal of Environmental Research and Public Health*, 6(11), 2880–2882. https://doi.org/10.3390/ijerph6112880

8. **Samet, J. M., Marbury, M. C., & Spengler, J. D.** (1987). Health Effects and Sources of Indoor Air Pollution. Part I. *American Review of Respiratory Disease*, 136(6), 1486–1508. https://doi.org/10.1164/ajrccm/136.6.1486.
9. **Whitmer, M.** (2025). The Air Inside Your Home May Actually Be More Polluted Than Outside, New Study Says. *Real Simple*. https://www.realsimple.com/indoor-air-quality-study-11682840
10. **Zhang, Y., et al.** (2020). Mitigation of Indoor Air Pollution: A Review of Recent Advances in Adsorption Materials and Catalytic Oxidation. *Science of the Total Environment*, 743, 140679. https://doi.org/10.1016/j.scitotenv.2020.140679

Climate Change Adaptation and Mitigate Measures for the Impact of Air Pollution in Indian Metropolitan City

AJAYA INDRANI * , **Shailendra Kumar Mandal, Kamini Sinha**
NATIONAL INSTITUTE OF TECHNOLOGY PATNA

Climate Change Adaptation and Mitigate Measures for the Impact of Air Pollution in Indian Metropolitan City

Ajaya Indrani*, Shailendra Mandal, Kamini Sinha

National Institute of Technology Patna, India

Department of Architecture and Planning NIT Patna, Patna, India

Corresponding author email: ajayai.pg23.ar@nitp.ac.in

ABSTRACT

This research examines air pollution and associated health hazards in the Indian Metropolitan city of Patna, which covers 1167.04 sq.km and has an estimated population of 3.88 million. As one of the world's most polluted cities, Patna faces significant air quality challenges exacerbated by its geographical location in the Indo-Gangetic Plain & unique meteorological conditions that trap pollutants.

The study examines air pollution in Patna through ground-based monitoring stations, satellite data (Sentinel-5P), and Google Earth Engine. Major pollutants analysed include PM2.5, PM10, SO2, NO2, and CO with analysing the primary pollution sources. Pollution level peaks during winters due to temperature inversions and biomass burning.

The research documents the study of topographical and land-use analysis concerns about agricultural land conversion to urban uses, with built-up areas showing the largest net gain while agricultural lands experienced the most significant losses.

Based on the findings, the study suggests the rudimentary measures related to the Patna's specific air quality adaptation and mitigation measures to improve the air quality of the Patna city. The findings emphasize the need for integrated approaches that address both local emissions and regional dynamics, offerings valuable insights for air quality management strategies in developing urban regions.

References:

CENTRAL POLLUTION CONTROL BOARD. (n.d.). STANDARDS. NATIONAL AIR QUALITY INDEX.
BIHAR STATE POLLUTION CONTROL BOARD. (n.d.). ACTION PLAN FOR CONTROL OF AIR POLLUTION IN NON-ATTAINMENT CITY OF BIHAR (PATNA).
BSPCB. (n.d.). (report). COMPREHENSIVE CLEAN AIR ACTION PLAN FOR THE CITY (Report).

Keyword: *Environmental Health, Air Pollutants, Metropolitan Area*

"A Side-by-Side Comparative study on change in effectiveness of Hybrid Air Cleaner when tested in accredited laboratory environment versus real site with centrally air conditioned space scenario for Tackling PM and Toxins"

KAPIL KAPOOR * [1], **Parvinder Kaur Kapoor**[2], **Avinash Kumar**[3]

[1] *CTO-Vayuguard Climate Tech Private Limited, Research & Innovation Park, IIT Delhi Campus, New Delhi, India*

[2] *CAO-Vayuguard Climate Tech Private Limited, Research & Innovation Park, IIT Delhi Campus, New Delhi, India*

[3] *CMO-Vayuguard Climate Tech Private Limited, Research & Innovation Park, IIT Delhi Campus, New Delhi, India*

This research paper examines the impact of Hybrid Electro-Magnetic or Hybrid Central Air Cleaner and Sanitizing (HCAC) technology on indoor air quality (IAQ) by focusing on the reduction of particulate matter (PM) concentrations in corporate office environments. Specifically, it evaluates the effectiveness of HCAC units in lowering levels of PM1, PM2.5, and PM10, which are known contributors to health complications related to poor IAQ. The study is conducted in both an accredited laboratory setting and a real-world centralized air-conditioned office, allowing for a comprehensive analysis of the air filtration system's performance. Additionally, the research considers various external factors that may influence filtration efficiency and includes the assessment of other pollutants such as Formaldehyde and Total Volatile Organic Compounds (TVOC). The findings aim to provide insights into the potential benefits of advanced air purification technologies in enhancing occupant health and well-being in indoor environments.

Keyword: *Hybrid Air Sanitization and Purification, Advanced air purification techniques, High efficiency air treatments*

Performance comparison of commercially available unipolar standalone ionisation devices in terms of particulate matter capture

Aiswarya Kumar * , **Prashant Nawale, Manoranjan Sahu Sahu**
IIT bombay

Submicron particulate matter is found to cause health concerns. Although several PM control technologies are available on the market and lab scale, they have different drawbacks. In this regard, ionization has multiple advantages. Even ionization devices such as ionizers as well as electrostatic precipitators (ESP) pose several drawbacks. Therefore, the current study adopted different carbon brush ionizers as charger modules and integrated them with the grounded collection module of ESP in a closed, well-defined device volume; this way, we could reduce some of the drawbacks of both technologies. The performance of six ionization devices, including four commercially available carbon brush ionizers, a customized spiked static bar and a stainless-steel discharge electrode, was evaluated with respect to the efficiency of PM capture from primary indoor sources as well as from real indoor scenarios. Ionizers were ranked as best-performing, low-performing, and poorly performing based on PM capture efficiency. Results from the study showed a V-shaped curve with total efficiency ranging from 10.47 to 99.02 % for sizes 10 nm to 1000 nm. The trade-off between ion generation rate and orientation was found to affect PM capture of tested ionizers. As ESP generated ultrafine PM effectively it was performing poorly in case of mosquito coil as well as mixed aerosol with a total efficiency of 75.68 and 86.46 % respectively. Therefore, it was integrated with the best-performing carbon brush ionizer I4, thereby allowing the hybridized ionization system to generate ions in both vertical and horizontal directions with maximized capture of ultrafine PM.

Keyword: *Carbon brush, submicron, static road, indoor environment, polarity, ionization, Carbon brush, Submicron PM*

Botanical biofilters degrade toxic indoor air pollutants

ALLAN AUGUSTO ALVARADO ALVARADO * [1], **SIEGFRIED DENYS**[1], **WENKE SMETS**[2]
[1] *Antwerp engineering, electrochemistry and sensing (A-PECS), Department of Bioscience Engineering, University of Antwerp, Belgium*
[2] *Laboratory of Applied Microbiology and Biotechnology (LAMB), Department of Bioscience Engineering, University of Antwerp, Belgium*

Indoor air quality (IAQ) has received less attention than outdoors, although people spend >90% of their time indoors. The concentration of indoor air pollutants like volatile organic compounds (VOCs) can be 2 - 5 times higher than outdoors, being highly detrimental to human health, particularly benzene and formaldehyde. The total VOC concentration exceeds the threshold limit in the Flemish Region of Belgium. Current technologies to abate indoor VOCs are predominantly physical-chemical methods with high energy consumption, pollutant phase-transfer, and the generation of hazardous by-products. This work scrutinizes the removal and biodegradation of VOCs employing active botanical biofiltration. A botanical biofilter possesses a wider range of mechanisms to remove VOCs, as the substrate, rhizosphere, and phyllosphere are inhabited by complex bacterial communities that utilize VOCs as a carbon source. Two lab-scale botanical biofilters were built using Epipremnum sp. and substrate (pine bark, expanded clay, compost) and acclimatized for 15 days (8 hours day^{-1}) in a 1.2 m^3 climate chamber to a mixture of priority VOCs i.e., benzene, toluene, ethylbenzene, xylene, (BTEX) acetaldehyde and formaldehyde at five times the allowable indoor concentration. The airflow rate was set at 60 m^{-3} h^{-1}, and the VOC concentration was monitored in real-time using innovative SIFT-MS (Selected Ion Flow Tube Mass Spectrometry). Kinetic models revealed the type of mechanisms responsible for the VOC removal, and the single-pass removal efficiency (SPRE) was derived. It was confirmed that the SPRE of BTEX increased daily until reaching a linear decay due to biodegradation, opposite to acetaldehyde and formaldehyde, which followed a first-order decay, likely due to adsorption. CO_2 production confirmed the biological removal of the VOCs and changed when varying the VOC initial concentration. Finally, the airflow rate was increased, leading to higher clean air delivery rates (CADR).

References:

Llewellyn, D., Dixon, M., 2011. Can plants really improve indoor air quality? Comprehensive Biotechnology, Second edition.,. Elsevier Inc., pp. 331–338. https://doi.org/10.1016/B978-0-08-088504-9.00325-1

Mikkonen, A., et al., 2018. Biofiltration of airborne VOCs with green wall systems—microbial and chemical dynamics. Indoor Air 28 (5), 697–707. https:// doi.org/10.1111/ina.12473.

Paull, N.J., Irga, P.J., Torpy, F.R., 2019. Active botanical biofiltration of air pollutants using Australian native plants. Air Qual Atmos Health 12 (12), 1427–1439. https://doi.org/10.1007/s11869-019-00758-w.

Keyword: *botanical biofiltration, volatile organic compounds (VOCs), bioremediation*

Air purification for removal of volatile organic compounds in a controlled indoor environment

Nigel Goodman * [1,2,7], **Max Rounds**[3], **Kevin Kevin**[3], **Martine Dennekamp**[4], **Jen Martin**[4], **Behzad Rismanchi**[5], **Jason Monty**[6], **Sotiris Vardoulakis**[1], **Amanda Wheeler**[1,7]

[1] *HEAL Global Research Centre, Health Research Institute, University of Canberra, Australian Capital Territory, Australia*

[2] *Sustainable Building Innovation Laboratory (SBi Lab), RMIT University, Melbourne, Victoria, Australia*

[3] *Department of Mechanical Engineering, The University of Melbourne, Victoria, Australia*

[4] *Environment Protection Authority (EPA) Victoria, Melbourne, Victoria, Australia*

[5] *Department of Infrastructure Engineering, The University of Melbourne, Victoria, Australia*

[6] *Electrical, Mechanical and Infrastructure Engineering, The University of Melbourne, Victoria, Australia*

[7] *Commonwealth Scientific Industrial Research Organisation (CSIRO) Environment, Melbourne, Victoria, Australia*

Background: In Australia, there is a legacy of environmental contamination by industrial chemicals. In the worst cases chemicals have been disposed of by pouring into drains or over land to evaporate impacting indoor air quality through hydrocarbon vapour intrusion (HVI). Volatile organic compounds (VOCs) (e.g., chlorinated hydrocarbons, benzene) are pollutants associated with HVI. Air cleaners that incorporate adsorbent media (e.g., activated carbon) can remove or reduce levels of volatile pollutants from (HVI).

Aims: This study evaluated the performance of a gas phase air cleaner under a range of operational conditions in a controlled indoor environment.

Methods: The evaluation was conducted according to a detailed sampling protocol that followed established methods (i.e., US EPA Compendium Method TO-17) for sampling and analysis of VOCs. The test room was configured to have similar ventilation rates observed in Australian residential buildings (i.e., 0.8–1.2 ACH).

Results: Under a range of operational scenarios (i.e., steady state, continuous recirculation, single pass), a reduction in the levels of VOCs to background chamber levels was consistently achieved. Evaluation of the single pass efficiency revealed that the air cleaner had capacity to remove between 40% and 66% of the individual or total volatile compounds from the room.

Implications: This project has demonstrated improvements for indoor air quality in terms of levels of VOCs within a simulated indoor environment. Further research to evaluate the long-term performance of the air filter would be a valuable addition to the project, as would testing the device in a building impacted by hydrocarbon vapour intrusion.

Keyword: *indoor air quality, hydrocarbon vapour intrusion, air purification*

Assessment and Characterization of Airborne Microplastics Across Indoor Settings

Name: Chaithra S, Indian Institute of Technology Madras
Contact Mail: chaithra.sraju7@gmail.com
Authors: Chaithra S a, S M Shiva Nagendra a*

Abstract:
Airborne microplastics have emerged as a significant concern in recent years, particularly within the indoor environments since individuals spend approximately 90 - 95 % of their time in indoors. Microplastic presence in indoor air, not only contributes to environmental degradation but also poses potential health risks to the occupants due to prolonged exposure. This study aims to comprehensively examine the occurrence, characterization and morphological features of airborne microplastics across different indoor environments, including a 3D printing room, waste management facility, ball mill, workshop, classrooms and corridor space and an incineration plant. The highest concentration of airborne microplastics was observed in the incineration plant (9.31 MPs/m^3), followed by classrooms and corridor spaces (6.35 MPs/m3), the waste management site (5.52 MPs/m^3) and the 3D printing room (5.51 MPs/m^3). These findings highlight the variation in microplastic concentration across different indoor environments. Particles smaller than 100 µm were the most predominant across all locations, with the highest proportion observed at the incineration plant (65%), followed by waste management facility and 3D printing room (both at 42%). Particle size ranging from 200 – 500 µm were the second most abundant, particularly in the ball mill and workshop (both at 19%). Particles larger than 1000 µm size were mainly found in classrooms and corridor space (22%), followed by department workshops (20%). Across all locations, most microplastic particles were found to be in blue, black and transparent color, with the majority being fibrous in shape, followed by fragment shaped particles. Polyacrylates, polyolefins, polyvinyl-based compounds, and polyesters were the most commonly identified polymer types across the observed indoor environments. This study reveals that, the generation and emission of microplastics are strongly influenced by the primary activities happening within these spaces, which may pose a significant health risk to the occupants upon prolonged exposure.

THE MULTIFACETED INFLUENCE OF INDOOR PLANTS ON INDOOR AIR QUALITY AND OCCUPANT WELL-BEING

Hind Rasheed * , **Jayasree T K**
National Institute of Technology Calicut

Indoor air pollution significantly impacts human health, comfort, and well-being, with individuals spending 80–90% of their time indoors. Increasing health issues underscore the importance of effective indoor air quality management strategies. Botanical biofilters are recognized as a cost-effective and energy-efficient solution for reducing indoor gaseous pollutants.

This study examines the multifaceted impacts of tropical plants, focusing on their pollutant sequestration capabilities and psychological benefits. A pilot survey was conducted following a comprehensive literature review to address key research questions regarding occupant interaction with plants and their perceptions.

Participants were categorized into three groups: those with plants at home, in workspaces and in both settings. Six subsections, including distinct groups of phytophiles and non-phytophiles, evaluated knowledge about common tropical plants, their characteristics, and pollutant-sequestering abilities. The findings revealed that Golden Pothos, Snake Plant and Aloe Vera are the most commonly preferred species and that plant selection is primarily influenced by aesthetics (81.3%), ease of maintenance (68.8%) and cost (50%). The integration of indoor plants is linked to increased productivity (68.6%) and creativity (42.9%), especially in hybrid home-office environments. Additionally, biophilic design principles emphasize plants' psychological benefits, such as stress reduction and enhanced cognitive function.

This study underscores the importance of transitioning from traditional chamber-based experiments to real-time monitoring of plant efficiencies for more accurate and practical assessments. By leveraging plant-based filtration systems, this study provides insights on the perceptions of users in integrating plants that foster healthier and more sustainable indoor environments.

Keyword: *Indoor plants, Pollutant sequestration, Indoor Air Quality*

The multifaceted influence of indoor plants on indoor air quality and occupant well-being

Hind Rasheed[1]*, Dr. Jayasree T K[1]

[1] Department of Architecture and Planning, National Institute of Technology Calicut, India
*hind_p230576ar@nitc.ac.in

SUMMARY

Indoor air pollution poses significant health risks, necessitating effective mitigation strategies. This study explores the dual functionality of indoor plants as botanical biofilters for air purification and as elements that enhance psychological well-being. By examining their pollutant sequestration capabilities and psychological benefits, this research provides insights into the role of tropical plants in improving indoor air quality. A comprehensive literature review was conducted, followed by a pilot survey assessing occupant interaction with plants. Participants were categorized into three groups based on plant presence at home, workspaces, or both, to explore plant knowledge, user preferences and pollutant-removal perceptions among phytophiles and non-phytophiles. The findings reveal that *Epipremnum aureum*, *Dracaena trifasciata* and *Aloe barbadensis miller* are the most preferred indoor plants, with selection primarily driven by aesthetics (81.3%), ease of maintenance (68.8%) and cost (50%). The integration of plants is linked to perceive increase in productivity (68.6%) and creativity (42.9%), particularly in hybrid home-office environments. By incorporating biophilic design principles, indoor spaces can promote well-being through stress reduction and cognitive benefits. Moreover, this study underscores the importance of transitioning from laboratory-based experiments to real-world applications, advocating for real-time monitoring of plant efficiencies to optimize their role in sustainable indoor air quality management.

KEYWORDS

Indoor plants; Pollutant sequestration; Indoor Air Quality; Botanical filtration; Plant perception

1 INTRODUCTION

Indoor air pollution represents one of the most significant environmental health concerns of the modern era, particularly considering that people spend approximately 80-90% of their time indoors (Bandehali et al., 2021; Liu et al., 2022; Mosaddegh et al., 2014; Pichlhöfer et al., 2021; Ravindra & Mor, 2022). The quality of indoor air directly impacts human health, comfort, and overall well-being, with poor air quality linked to a range of health issues from minor irritations to serious respiratory conditions (Ibrahim & Chong, 2019). Exposure to indoor air pollution has been linked to various health risks, including respiratory illnesses and sick building syndrome (Al horr et al., 2016). In India, indoor air pollution is a critical issue, contributing to approximately 18.5% of deaths annually (IHME, 2024).

Indoor landscaping has emerged as an environmental friendly approach to reducing airborne contaminants. Biophilic design, which incorporates natural elements into built environments, has gained recognition for its potential to enhance human well-being and environmental sustainability. Indoor plants serve as a central component of biophilic design strategies, fulfilling both aesthetic and functional roles. However, research on the real-world effectiveness of plants in improving IAQ remains limited (Bedi & Bhattacharya, 2021). This study comprises of two components: a systematic review of existing literature on the role of indoor plants in air purification and an analysis of plant-people interaction perceptions based on a pilot study.

Plant Mechanisms for Indoor Air Pollution Reduction

Research on plants' ability to improve indoor air quality gained prominence following the landmark NASA Clean Air Study in the late 1980s (Thomas et al., 2015), which identified numerous indoor plants capable of removing common indoor air pollutants. Subsequent studies have expanded this foundation, investigating the specific mechanisms through which plants filter air pollutants and quantifying their effectiveness in various indoor environments (Etim, 2012; Ibrahim & Chong, 2019; Jhanji & Dhatt, 2021; Ravindra & Mor, 2022).

Plants remove air pollutants through various mechanisms, including absorption through leaf stomata, where gases enter during normal gas exchange and adsorption to plant surfaces, where particles adhere to waxy cuticles on leaves. Additionally, pollutants can be deposited in the root zone, where soil microorganisms break down harmful compounds or metabolized within plant tissues. Different plant species exhibit varying capabilities in removing specific pollutants, making them valuable natural air purifiers for indoor environments (Ibrahim & Chong, 2019; Mannan & Al-Ghamdi, 2021).

Among the most effective air-purifying plants, *Epipremnum aureum* (Golden Pothos) efficiently removes benzene, formaldehyde, xylene and other toxins thriving in low-light conditions with minimal care (Etim, 2012; Gong et al., 2019; Pichlhöfer et al., 2021). *Dracaena trifasciata* (Snake plant) known for its resilience, filters multiple air pollutants while requiring little maintenance. *Aloe barbadensis miller* (Aloe vera) eliminates formaldehyde and releases oxygen at night, making it ideal for bedrooms (Jhanji & Dhatt, 2021). *Spathiphyllum spp.* (Peace Lily) effectively removes ammonia, formaldehyde and benzene while enhancing aesthetics with its white flowers. *Chlorophytum comosum* (Spider plant) is a low-maintenance option that removes carbon monoxide and formaldehyde (Jhanji & Dhatt, 2021). *Chamaedorea seifrizii* (Bamboo Palm) and *Nephrolepis exaltata* (Boston fern) filter formaldehyde and other toxins while adding humidity to indoor spaces. *Dypsis lutescens* (Areca Palm) highly ranked for pollution removal, effectively eliminates xylene and toluene contributing to healthier indoor environments. The efficacy of plant-based air purification depends on various factors, including pollutant type and concentration, plant species and size, growing media composition and environmental conditions (Rasheed & Jayasree, 2025).

Psychological and Physiological Benefits of Indoor Plants

Indoor plants substantially enhance human well-being in addition to their air-purifying capabilities (Rasheed & Jayasree, 2025). Research has demonstrated significant correlations between exposure to indoor plants and greater mental health, less stress and improved cognitive performance. A cross-sectional study among urban Chinese adults revealed that persons identifying as 'houseplant caretakers' had superior mental well-being compared to 'non-houseplant carers'. The study demonstrated positive associations between mental well-being and variables like time dedicated to plant care, quantity of indoor plants maintained and duration of plant-keeping experience (Ma, 2022). The psychological benefits of indoor plants may be explained by the biophilia hypothesis, which suggests an innate human tendency to connect with nature. This connection with natural elements even in indoor settings can provide restoration from mental fatigue and stress. Research indicates that exposure to nature improves attention span and helps cope with attention deficit (Chamola et al., 2022).

Based on physiological perspective, indoor plants influence various human functions. Studies have shown that the presence of plants in hospital environments is associated with more stable vital signs among patients, including blood pressure, heart rate, respiration rate and body temperature (Sugano et al., 2022). Plants also contribute to improved indoor humidity levels and air cooling enhancing physical comfort in indoor spaces (Mannan & Al-Ghamdi, 2021). In workplaces, the presence of plants is associated with enhanced productivity, creativity, and job satisfaction. Research indicates that employees in plant-rich environments exhibit greater productivity compared to those in plant-free spaces. Likewise, students in classrooms adorned with plants experience improved concentration and psychological relaxation. The domestic sphere also reaps benefits from the incorporation of plants as indoor gardens are deemed therapeutic for mental health. Studies reveal that the lack of plants in indoor environments can result in irritation and hostility, underscoring the detrimental effects of disconnection from natural elements. (Chamola et al., 2022).

2 METHODS

This study employed a mixed-methods approach, combining a comprehensive literature review with a pilot survey to investigate the multifaceted roles of indoor plants. The literature review synthesized existing research on plants' air-purifying capabilities, psychological and applications in biophilic design. The pilot survey gathered primary data on user perceptions, preferences and experiences with indoor plants in various settings.

Survey Development and Implementation

The structured questionnaire developed on the basis of research questions formulated during the literature review, included both closed and open-ended questions to capture nuanced responses and was administered online via Google Forms to reach a diverse participant pool.

Participant Categorization and Analysis Framework

Respondents were categorized into three primary groups based on plant presence: i. Individuals with plants at home, ii. at the workplace and iii. in both settings. Further classification into six subsections distinguished phytophiles from non-phytophiles based on self-reported attitudes and plant-keeping behaviours enabling an analysis of plant affinity influence on perceptions and experiences. The questionnaire followed a systematic structure to align with research objectives. The first section gathered demographic data. The second examined plant ownership, type and selection factors. The third assessed awareness of indoor air quality and associated health effects. The fourth investigated knowledge of plant-based air purification, particularly species known for pollutant sequestration. The fifth analysed perceived benefits of indoor plants on productivity, creativity, psychological well-being and stress reduction. The final section addressed challenges in plant maintenance. This structured approach provided a comprehensive understanding of participants' engagement with indoor plants and their role in enhancing indoor environments.

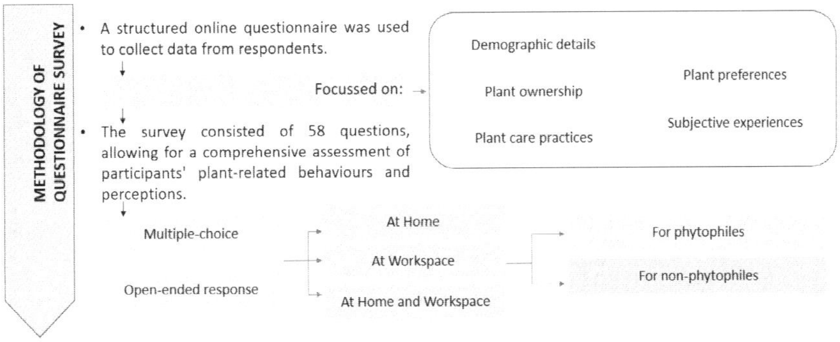

Figure 1 Methodological framework of questionnaire survey

3 RESULTS

The pilot survey gathered responses from participants representing diverse age groups, professional backgrounds, and living arrangements. This distribution allowed for comparative analysis of perceptions and experiences across different environments. The preferences, Figure 2, and selection criteria, Figure 3, varied across participant categories. Phytophiles demonstrated greater knowledge of plant species and their specific benefits, often selecting plants based on both aesthetic and functional considerations. Non-phytophiles prioritized low-maintenance requirements and aesthetic appeal in their selections such as *Golden pothos*, with less emphasis on specific air-purifying functions.

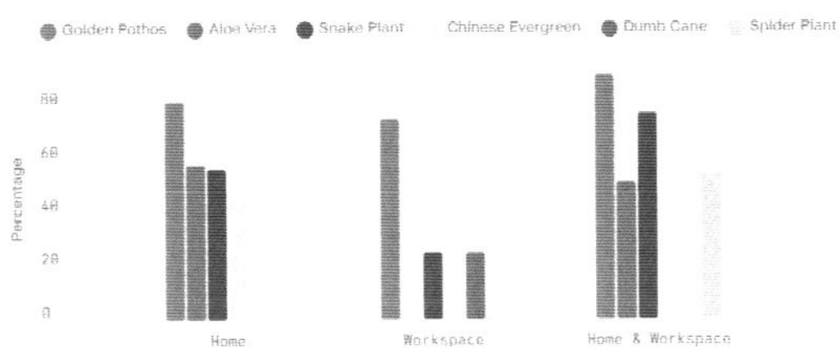

Figure 2 Most preferred indoor plants

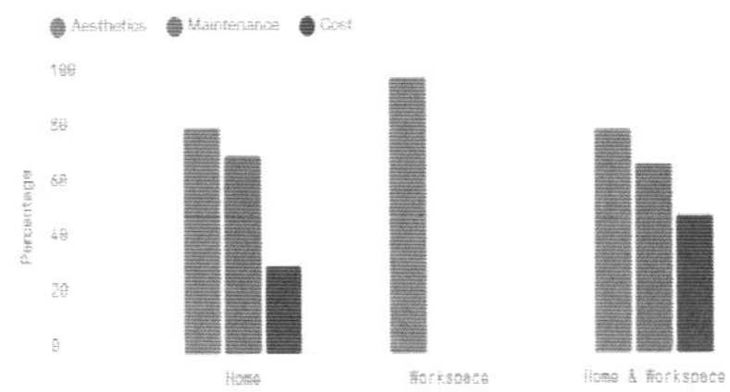

Figure 3 User criteria for plant preference

The most comprehensive benefits were reported by participants with plants in both home and work environments. This group reported higher overall productivity graph, Figure 7 and greater awareness of the multiple functions that plants serve in indoor spaces. Additionally, this group demonstrated the highest levels of engagement with plant care and maintenance. The survey results indicated that plants' psychological benefits were often perceived more immediately and strongly than their air-purifying functions. While many participants expressed awareness of plants' potential to improve air quality, the tangible, day-to-day benefits in terms of mood enhancement, stress reduction, and aesthetic improvement were more frequently cited as primary motivations for plant ownership.

The preference for 'combination' species is significantly higher compared to 'same type' species in all categories (Home, Workspace, and Home & Workspace). The observed differences in preferences are not due to random chance but reflect a meaningful trend in the survey data.

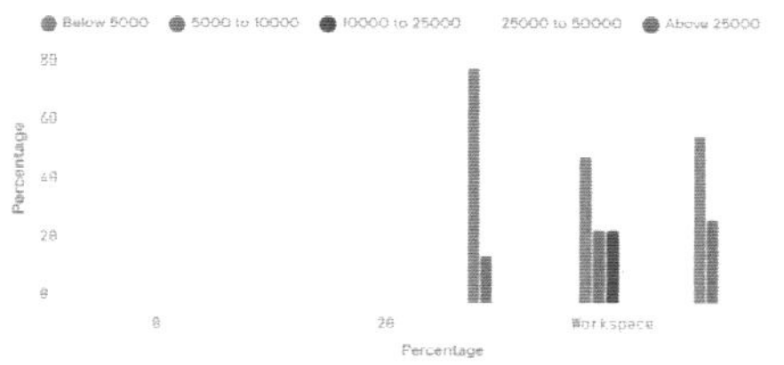

Figure 4 Amount spent for plant procurement and maintenance

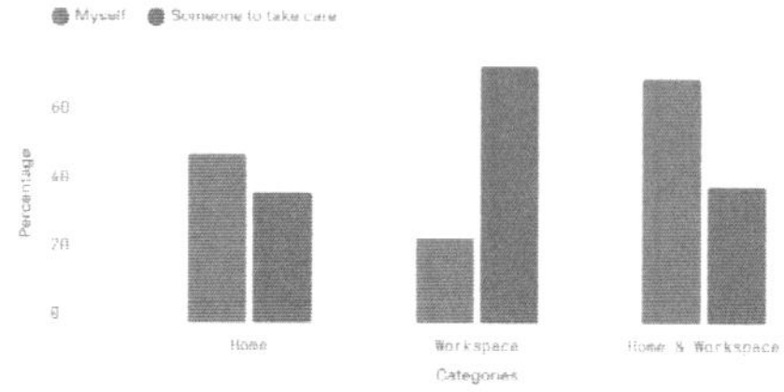

Figure 5 Plant care ritual

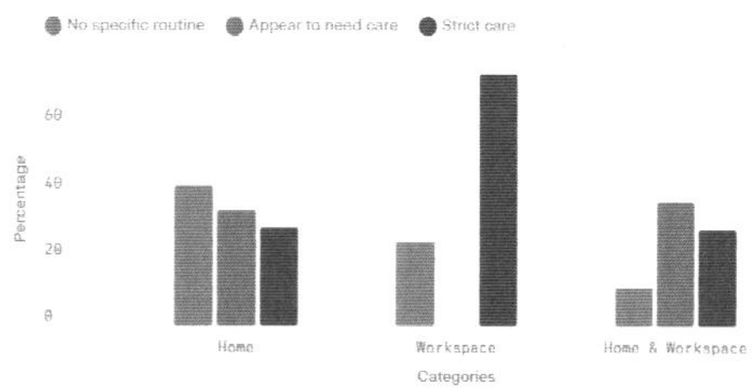

Figure 6 Plant care routine

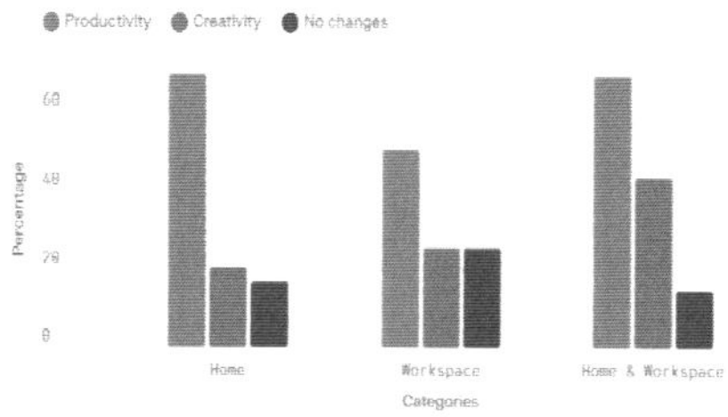

Figure 7 Impacts in the presence of indoor plants

4 DISCUSSION AND CONCLUSION

The findings of this pilot study underscore the significant role of indoor plants in enhancing air quality and occupant well-being. Integrating plants into residential and commercial spaces offers a cost-effective strategy for improving indoor environmental quality while balancing functional and aesthetic considerations. Increased productivity, creativity, and psychological comfort among participants highlight the broader human-centric benefits of plant installations.

Practical recommendations derived from the study emphasize selecting multi-benefit species strategic placement near pollution sources or high-traffic areas and combining species with complementary air-purifying capabilities. Additionally, aesthetic appeal plays a crucial role in plant acceptance and maintenance, necessitating a balance between function and form. Providing education on plant benefits can further enhance engagement, particularly among non-phytophiles. For shared spaces, a mix of low-maintenance common area plants and personalized plant programs may optimize benefits for diverse occupants. Despite its valuable insights, this pilot study has limitations, including a small sample size and potential self-reporting biases. Future research should expand sample diversity, employ real-time air quality monitoring, and conduct longitudinal studies to assess evolving plant-human interactions. Understanding optimal plant densities and species combinations across different indoor settings will further refine plant-based solutions for healthier environments.

As indoor occupancy continues to rise, creating restorative, sustainable environments become increasingly essential. This study highlights the dual benefits of indoor plants air purification and psychological well-being, reinforcing their potential as a viable strategy for healthier indoor spaces. By integrating scientific and human-centred approaches we can harness the natural capabilities of plants to enhance both environmental quality and human health paving the way for more sustainable and supportive built environments.

7 REFERENCES

Al horr, Y., Arif, M., Katafygiotou, M., Mazroei, A., Kaushik, A., & Elsarrag, E. (2016). Impact of indoor environmental quality on occupant well-being and comfort: A review of the literature. *International Journal of Sustainable Built Environment*, *5*(1), 1–11. https://doi.org/10.1016/j.ijsbe.2016.03.006

Bandehali, S., Miri, T., Onyeaka, H., & Kumar, P. (2021). Current state of indoor air phytoremediation using potted plants and green walls. *Atmosphere*, *12*(4). https://doi.org/10.3390/atmos12040473

Bedi, T. K., & Bhattacharya, S. P. (2021). An Investigative Study on Perceived Indoor Air Quality During COVID-19 Lockdown in India. *Journal of The Institution of Engineers (India): Series A*, *102*(4), 885–900. https://doi.org/10.1007/s40030-021-00546-9

Chamola, A., Apurva, S., Rao, A., & Ghosh, C. (2022). *Psychological and Physiological Impact of Indoor Plants and Their Maintenance*. *10*(3). https://doi.org/10.25215/1003.147

Etim, E. E. (2012). Phytoremediation and Its Mechanisms: A Review. *International Journal of Environment and Bioenergy International Journal of Environment and Bioenergy Journal Homepage: Www.ModernScientificPress.Com*, *2*(3), 120–136.

Gong, Y., Zhou, T., Wang, P., Lin, Y., Zheng, R., Zhao, Y., & Xu, B. (2019). Fundamentals of ornamental plants in removing benzene in indoor air. *Atmosphere*, *10*(4). https://doi.org/10.3390/ATMOS10040221

Ibrahim, I. Z., & Chong, W. T. (2019). *I ndoor and Built Evaluation of common indoor air pollutant reduction by a botanical indoor air biofilter system. November*. https://doi.org/10.1177/1420326X19882080

IHME. (2024). *Our World in Data*. Global Burden of Disease. https://ourworldindata.org/air-pollution

Jhanji, S., & Dhatt, U. K. (2021). Phytoremediation of indoor air pollutants: Harnessing the potential of plants beyond aesthetics. *Journal of Horticultural Sciences*, *16*(2), 131–143. https://doi.org/10.24154/jhs.v16i2.986

Liu, F., Yan, L., Meng, X., & Zhang, C. (2022). A review on indoor green plants employed to improve indoor environment. *Journal of Building Engineering*, *53*(March), 104542. https://doi.org/10.1016/j.jobe.2022.104542

Ma, J. (2022). *Interaction with Nature Indoor : Psychological Impacts of Houseplants Care Behaviour on Mental Well-Being and Mindfulness in Chinese Adults*.

Mannan, M., & Al-Ghamdi, S. G. (2021). Active Botanical Biofiltration in Built Environment to Maintain Indoor Air Quality. *Frontiers in Built Environment*, *7*(June), 1–8. https://doi.org/10.3389/fbuil.2021.672102

Mosaddegh, M. H., Jafarian, A., Ghasemi, A., & Mosaddegh, A. (2014). Phytoremediation of benzene, toluene, ethylbenzene and xylene contaminated air by D. deremensis and O. microdasys plants. *Journal of Environmental Health Science and Engineering*, *12*(1). https://doi.org/10.1186/2052-336X-12-39

Pichlhöfer, A., Sesto, E., Hollands, J., & Korjenic, A. (2021). Health-related benefits of different indoor plant species in a school setting. *Sustainability (Switzerland)*, *13*(17). https://doi.org/10.3390/su13179566

Rasheed, H., & Jayasree, T. K. (2025). Energy & Buildings The multifaceted role of indoor plants : A comprehensive review of their impact on air quality , health , and perception ☆. *Energy & Buildings*, *330*(September 2024), 115312. https://doi.org/10.1016/j.enbuild.2025.115312

Ravindra, K., & Mor, S. (2022). Phytoremediation potential of indoor plants in reducing air pollutants. *Frontiers in Sustainable Cities*, *4*. https://doi.org/10.3389/frsc.2022.1039710

Sugano, S., Tazaki, M., Arai, H., Matsuo, K., & Tanabe, S. I. (2022). Effects of Indoor Plants on Occupants' Cognitive Functions: A Systematic Review. *17th International Conference on Indoor Air Quality and Climate, INDOOR AIR 2022*.

Thomas, C. K., Jin Kim, K., & Kays, S. J. (2015). Phytoremediation of Indoor Air 765 MISCELLANEOUS. *Hortscience*, *50*(5), 765–768.

Optimization of Affordable and Durable Retrofitting Solutions for Vernacular Buildings in Rural India: A Multi-Objective Approach

Dharavath Jaicharan * [1], **Jayasree TK**[1], **Deepthi Bendi**[1], **Neelima Geetha Archana**[2]
[1] *Department of Architecture and planning, NIT Calicut*
[2] *Arizona State University, USA*

Vernacular buildings in rural India reflect traditional construction and cultural heritage but suffer from poor durability, lack of amenities, and environmental vulnerabilities. Socio-economic constraints and limited access to government initiatives leave these communities unable to afford modern retrofitting. This highlights the urgent need for affordable, durable, and sustainable retrofitting strategies, with a focus on optimizing retrofitting solutions for vernacular buildings in rural India. The study addresses critical objectives of economic affordability, environmental sustainability, and durability. By utilizing a multi-objective approach within resource-constrained boundaries, the research aims to identify retrofit strategies that balance cost-effectiveness, environmental impact, and extended lifespan. Key research questions include: How can locally available resources be effectively utilized for retrofitting solutions? What are the alternative options for retrofitting, and how can conflicts between objectives such as cost and durability be addressed? By employing advanced optimization techniques, such as Pareto-based methods, and integrating life-cycle assessments, the framework is expected to evaluate potential solutions under varying constraints. The findings will emphasize the importance of affordable and sustainable retrofitting measures in preserving vernacular architecture while enhancing resilience, making them applicable for large-scale rural development initiatives.

Keyword: *Residential dwellings, Retrofitting, Multi-objective optimization*

Thermal Envelope Performance of Commercial Space in a High Rise and High Density Setting

Tirth Shah * , **Krishan Upadhyay**
Institute of Architecture and Planning, Nirma University

This paper investigates the thermal performance of commercial office spaces in high-rise, high-density urban environments, with a focus on the impact of microclimates on building envelopes in subtropical climates. The rapid urbanization and population growth in metropolitan areas have led to an increased prevalence of urban heat island effects and surface heating, which exacerbate indoor thermal discomfort and elevate energy consumption in commercial buildings. However, the thermal response of buildings in such settings is often overlooked during design and development phases.

The research employs a hybrid simulation approach to assess the thermal behavior of a base case, shoebox model, and an existing case study. The aim is to analyze the influence of various building envelope parameters, such as orientation, shape, and material composition, on the overall thermal performance. By simulating the building's response to microclimate variations in high-density urban areas, this study seeks to provide evidence-based design guidelines for optimizing thermal envelopes, ensuring occupant comfort, and reducing energy demands in commercial developments.

Focusing specifically on the Indian context, the research addresses the unique challenges posed by high-rise, high-density settings in subtropical climates. The expected outcome includes a comprehensive framework for optimizing building envelopes, providing design recommendations for composite assemblies, and proposing strategies for mitigating the urban heat island effect. Ultimately, the findings aim to contribute to the development of sustainable solutions for commercial buildings in rapidly growing urban areas, where high-density living and migration are driving demand for more energy-efficient, thermally responsive structures.

Keyword: *Microclimate, Thermal Performance, Passive Survivability*

Development of a Deep Learning Algorithm for Size Classification of Indoor Microplastics

Sruthy Robert[1], Sonal Ayyappan[1] , Ratish Menon[1*]

1. SCMS School of Engineering and Technology, Karukutty, Kerala, India
Affiliated to APJ Abdul Kalam Technological University,Thiruvananthapuram, Kerala,India
2. *Corresponding email: ratishmenon@scmsgroup.org*

SUMMARY

One of the major challenges in studying indoor microplastics, especially in developing countries is the time-consuming and expensive process of particle identification and characterization. This study presents an automated deep learning approach using a modified U-Net Architecture, a fully convolutional network to segment and classify microplastics from microscopic images. U-Net combines both downsampling for contextual feature extraction with upsampling for precise localization making it effective even with limited training data. To overcome data scarcity, the study has employed extensive image augmentation techniques to improve the model's robustness. The model achieved a testing accuracy of 65% and an Intersection over Union (IoU) score of 0.48. The model was also able to provide quantifiable features into microplastic properties such as equivalent diameter and area, which is crucial for assessments and source tracking. Thus, this approach offers a practical step toward automated indoor microplastic monitoring and effective risk management.

KEYWORDS
U-Net, Fully Convolutional Network, Up sampling, Down Sampling, Automated Classification

1 INTRODUCTION

The conventional method of microplastic classification is a tedious process and will have errors. In this paper, a better and more efficient method which is also convenient is explored. The classification of microplastics is done by using an extension of a Fully Convolutional Network, the U-Net Architecture. In this mode, the desired output is localised which means that a class label is supposed to be assigned to each pixel. Moreover, thousands of training images are usually beyond reach in classification tasks. The architecture is much more efficient and productive than CNN-based systems, and it proposes a classifier output that considers the features of multiple layers. This work is closely related to (Ronneberger et al., 2015)

where a network and training strategy relies on the strong use of data augmentation to use the available annotated sample more efficiently. The architecture consists of a contracting path to capture context and a symmetric expanding path that enables precise localization. They show that such a network can be trained end to end from very few images and outperforms the prior best method on the ISBI challenge for segmentation of neuronal structure in electron microscopic stacks. Likewise, (Wollmann et al., 2019) proposed a new method for particle detection, which uses deep learning and is based on a domain-adapted Deconvolution Network approach on the ISBI Particle Tracking Challenge data and live cell fluorescence microscopy data of hepatitis C virus proteins. It was found that their approach produces high detection and localization performance for particles of different shapes and also for various signal-to-noise ratios.

2 METHODS

2.1 Sample Collection and Image Acquisition

Indoor microplastic particles were collected from common deposition and accumulation zones in residential indoor environments. The microplastic samples were cleaned as per (Prata et al., 2020; S et al., 2022) to remove any organic matter and were then filtered onto Whatman filter paper (Grade 42 with a pore size of 2.5µm). High-resolution microscopic images were captured using Magnus MagCam (5 MP) camera at 20X and 40X magnification using a stereo microscope, which provided necessary data for subsequent segmentation and classification. The dataset was expanded by 50% by augmentations for the model to become more robust against diverse shapes, sizes, and orientations of indoor microplastics.

2.2 System Architecture

2.2.1 Fully Convolutional Networks (FCNs)

In this study, Fully Convolutional Networks (FCNs) form the foundation of the segmentation framework. FCNs are composed entirely of locally connected layers such as convolutions, pooling, and upsampling, without any fully connected (dense) layers.

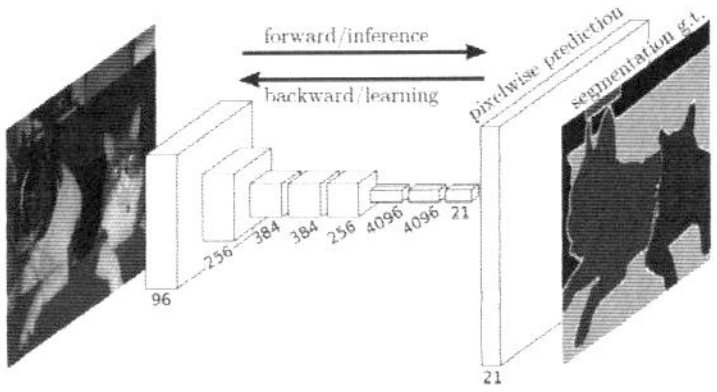

FCN Architecture (Long et al., 2015)

This design considerably reduces the number of trainable parameters and computational complexity, making the architecture efficient and adaptable to input images of varying dimensions (Long et al., 2015). The FCN architecture is structured into two primary components: a down-sampling path and an up-sampling path. The down-sampling path is responsible for capturing high levels of semantic and contextual information essentially understanding what is present in the image. On the other hand, the up-sampling path reconstructs spatial details to determine where specific features are located, enabling accurate segmentation. To enhance spatial precision, skip connections are introduced between corresponding layers in the down-sampling and up-sampling paths. These connections perform element-wise summation and serve to reintroduce fine-grained local information lost during pooling operations. The asymmetric structure of architecture with a reduced number of feature maps in the up-sampling path secures efficient computation while maintaining high segmentation fidelity. This structure makes FCNs a perfect fit for pixel-wise prediction tasks, such as microplastic detection in microscopy images (Huang et al., 2022).

2.2.2 U Net Architecture

In this study, we employ a modified five-layer U-Net architecture, an advanced form of Fully Convolutional Networks (FCNs), to achieve accurate microplastic segmentation from microscopic images. Compared to conventional FCNs, U-Net features a symmetric design with a contracting path that captures semantic context and an expansive path for precise localization. The key aspect of the effectiveness here is the skip connections that concatenate feature maps from corresponding layers, preserving fine-grained spatial information. A notable enhancement in implementation is the usage of a high number of feature channels in the expansive path. This design allows the contextual information captured during downsampling to be effectively propagated to higher-resolution layers improving localization accuracy. The contracting path comprises repeated applications of two unpadded 3×3 convolutions, each followed by a Rectified Linear Unit (ReLU) and a 2×2 max pooling operation with a stride of 2. At each pooling step, the number of feature channels is doubled. The expansive path includes upsampling the feature maps followed by a 2×2 convolution (up-convolution) that reduces the number of feature channels. This is concatenated with the corresponding cropped feature map from the contracting path, followed by two 3×3 convolutions and ReLU activations. Cropping is essential to align feature maps due to the loss of border pixels from unpadded convolutions. At the final layer, a 1x1 convolution is used to map each 64-channel feature vector to the target number of classes. The complete U-Net contains 23 convolutional layers. To handle large input images beyond GPU memory limitations, in this study an overlap-tile strategy is implemented with mirroring at the borders. This ensures full context is available at the edges, enabling accurate segmentation across the entire image.

For optimizing the model, the loss function used in this study is the Intersection Over Union (IOU), also known as Jaccard Index which is a measurement for determining the overlap between two areas. IOU will always be a value between 0 and 1. IoU measures both the size and spatial alignment of the regions: if two areas are identical in shape and location, IoU equals 1; if they don't overlap at all, IoU is 0. Even if shapes are similar, a mismatch in size or location leads to a lower IoU score, making it a reliable metric for evaluating segmentation accuracy in image analysis tasks. (Long et al., 2015; Wollmann et al., 2019, Sustersic et al., 2022).

3 RESULTS

3.1 Training Results

From the training results a graph was plotted for the validation and training loss over 1000 epochs (Figure 3.1 a). Both losses show a consistent downward trend, indicating effective learning by the model but in the later epochs, minor fluctuations in validation loss can be observed. But altogether, a validation accuracy and loss of about 68% and -69% respectively were obtained. The Jaccard Coefficient (IoU) for both training and validation data shows a steady improvement across epochs, eventually stabilizing around 70% for training and approximately 66% for validation (Figure 3.1 b).

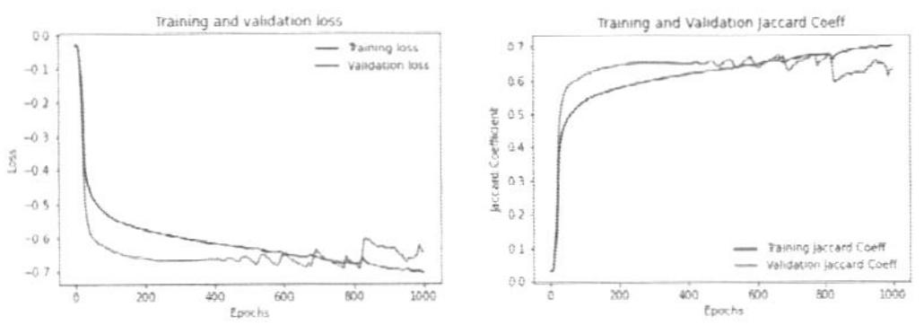

a) b)

Figure 3.1 (a) Training and Validation loss (b) Training and Jaccard Coefficient

3.2 Testing Results

During testing, the model achieved an IoU score of approximately 0.48 indicating moderate overlap. The given images after converting to binary images the model could identify the boundary regions and the structures with respect to the actual data (Figure 3.2).

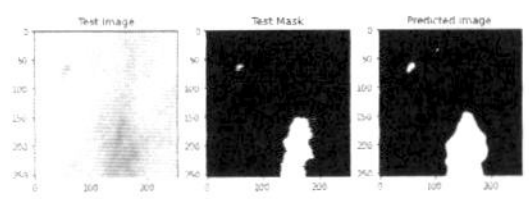

a)

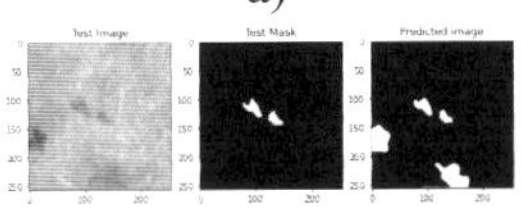

b)

Figure 3.2 (a) Testing Plot 1 (b) Testing Plot 2 (c) IoU Score

3.3 Prediction Results

The predicted segmentation tasks show that the model was able to successfully localize and extract relevant features effectively and efficiently capture the pattern in the image. Considering the microplastic pieces to their equivalent diameter the graph is plotted with the count of microplastics (Figure 3.3 - 3.4). The microplastic size range was below 140 micro metre and its corresponding count is also obtained by the model (Figure 3.5).

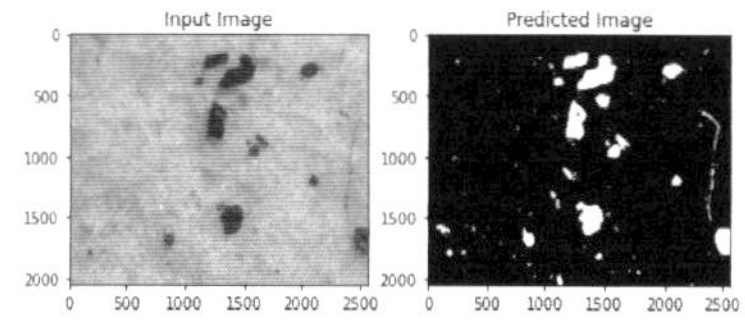

Figure 3.3 Predicted image with its corresponding input image

a) b)

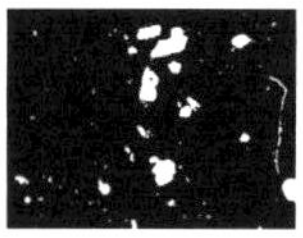

Figure 3.4 (a) Predicted Binary image (b) Instance Segmentation of the predicted binary image using Watershed algorithm

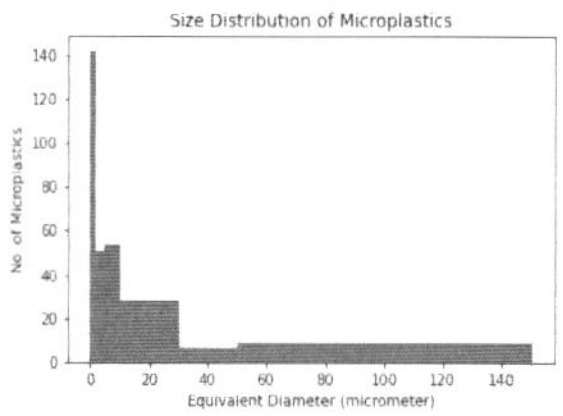

Figure 3.5 Size Distribution of MPs with respect to Equivalent Diameter

4 DISCUSSIONS

The training results suggest that the model achieves sufficient stable performance with the given data. The close convergence of the curves in the training results indicates that the model not only learns effectively but also maintains good generalization performance without significant overfitting. Comparing with a similar study conducted by Lorenzo-Navarro et al. 2021 achieved a higher IoU (0.8) and classification accuracy (98.11%) on larger particles (1 to 5 mm) with images taken from the mobile camera using U-Net for segmentation followed by VGG16. The present study model showed moderate performance with a validation IoU of 66% and test IoU of 0.48. The present study focused on much smaller and heterogenous particles which is likely to be found in the indoor environment.

Considering the limitations such as lower-resolution imaging from a microscope camera, our model achieved a validation IoU of 66% and a test IoU of 0.48, which highlights a reasonable performance. Advancements in the current model with more training quality data can boost the accuracy of the model for providing more insights into the indoor environment.

5 CONCLUSIONS

This study demonstrates the potential of a deep learning-based approach for detecting and classifying indoor microplastics, which are smaller and harder to identify unlike those from other sources. Utilizing a Fully Convolutional Network and a five-layer U-Net architecture, microplastics were successfully segmented from microscope images. The model achieved a testing accuracy of 65% with an IoU score of 0.48, validating its effectiveness in distinguishing microplastics from complex backgrounds. A key strength of this method lies in its ability to extract quantitative metrics such as particle count, area, and equivalent diameter which is critical for assessing exposure levels and associated health risks. Additionally, the inclusion of size distribution analysis enhances interpretability, offering insights into microplastic variability in indoor environments. Even though the identification of different categories of microplastic forms is possible through more developments more information can be obtained such as the classification of microplastics into categories (fibers, foams, fragments, etc.) based on shape and color, and improving accuracy by deepening the U-Net architecture. With further refinement, this tool has the potential to aid in indoor air quality assessment, health studies, and building design improvements, marking a significant step toward data-driven microplastic exposure management.

6 ACKNOWLEDGEMENT

We would like to express our sincere gratitude to the Directorate of Climate Change (DoECC), Thiruvananthapuram, Government of Kerala for providing financial support and the SCMS Group of Institutions for providing this opportunity. I would like to also thank Mr. Jithin Nettikadan, Ms. Aparna, Ms. Farzana, Ms. Ardra Nair for their assistance.

7 REFERENCES

Huang, Sheng-Yao, Wen-Lin Hsu, Ren-Jun Hsu, and Dai-Wei Liu. 2022. "Fully Convolutional Network for the Semantic Segmentation of Medical Images: A Survey." *Diagnostics* 12 (11): 2765. https://doi.org/10.3390/diagnostics12112765.

Long, Jonathan, Evan Shelhamer, and Trevor Darrell. 2015. "Fully Convolutional Networks for Semantic Segmentation." Openaccess.thecvf.com. 2015. https://openaccess.thecvf.com/content_cvpr_2015/html/Long_Fully_Convolutional_Networks_2015_CVPR_paper.html.

Lorenzo-Navarro, Javier, Modesto Castrillón-Santana, Elena Sánchez-Nielsen, Borja Zarco, Alicia Herrera, Ico Martínez, and May Gómez. 2021. "Deep Learning Approach for Automatic Microplastics Counting and Classification." *Science of the Total Environment* 765 (April): 142728. https://doi.org/10.1016/j.scitotenv.2020.142728.

Prata, Joana C., Joana L. Castro, João P. da Costa, Armando C. Duarte, Mário Cerqueira, and Teresa Rocha-Santos. 2020. "An Easy Method for Processing and Identification of Natural and Synthetic

Microfibers and Microplastics in Indoor and Outdoor Air." *MethodsX* 7: 100762. https://doi.org/10.1016/j.mex.2019.11.032.

Ronneberger, Olaf, Philipp Fischer, and Thomas Brox. 2015. "U-Net: Convolutional Networks for Biomedical Image Segmentation." *Lecture Notes in Computer Science* 9351: 234–41. https:// doi.org/10.1007/978-3-319-24574-4_28.

S, Anupama, Nisha Luckins, Ratish Menon, Sruthy Robert, and Thushara Sudheish Kumbalaparambi. 2022. "Micro-Plastics in the Vicinity of an Urban Solid Waste Management Facility in India: Assessment and Policy Implications." *Bulletin of Environmental Contamination and Toxicology* 109 (6): 956–61. https://doi.org/10.1007/s00128-022-03588-2.

Sustersic, Tijana, Vesna Rankovic, Vladimir Milovanovic, Vojin Kovacevic, Lukas Rasulic, and Nenad Filipovic. 2022. "A Deep Learning Model for Automatic Detection and Classification of Disc Herniation in Magnetic Resonance Images." *IEEE Journal of Biomedical and Health Informatics* 26 (12): 6036–46. https://doi.org/10.1109/jbhi.2022.3209585.

Wollmann, T, C Ritter, J N Dohrke, J.-Y Lee, R. Bartenschlager, and K Rohr. 2019. "Detnet: Deep Neural Network for Particle Detection in Fluorescence Microscopy Images." *2022 IEEE 19th International Symposium on Biomedical Imaging (ISBI)*, April, 517–20. https://doi.org/10.1109/ isbi.2019.8759234

Impact of household ventilation on unequal exposure to cooking - a model analysis on rural buildings

Mrinmoy Chakraborty * , **Milind Kandlikar, Naomi Zimmerman, Amanda Giang**
University of British Columbia

Rural and economically challenged families often experience elevated levels of air pollution due to the continuous use of solid biomass fuels (SBFs) such as dung cakes, firewood, or agricultural waste for daily activities, such as cooking. The effects of indoor cooking emissions are exacerbated by confined spaces and inadequate building ventilation, which are critical to determining indoor pollutant concentrations and their dispersion. Economically disadvantaged households often overlook proper ventilation. Given that humans spend most of their time indoors, exposure to indoor pollutants largely depends on the duration for which individuals remain inside, particularly during cooking times. This research aimed to investigate the influence of indoor cooking emissions on daily exposure in various exposure groups by varying fuel usage and adjusting ventilation. We used a nested mass balance box model (NMB2), which simulates temporal changes in BC concentration, a pollutant generated from the incomplete combustion of SBFs. This model accounts for pollutant emission and transport resulting from the dynamics of the household and community levels, with local air quality and weather conditions. We simulated various scenarios combining household fuel use and ventilation ranges. To assess individual exposures, we calculated gender specific 24-hour BC exposure ratios (female-to-indoor and male-to-indoor), highlighting how proximity to pollution sources and indoor presence affect exposure. For households with low ventilation and SBF users, the female-to-indoor ratio was 0.32, compared to 0.22 for men. Improved ventilation increased these to 0.37 and 0.29, respectively, showing that although indoor BC levels decrease, personal exposure may not, especially for primary cooks. Enhanced ventilation can transfer emissions outside, increasing exposure in nearby areas. This complexity indicates that ventilation alone isn't enough; source control and emission dispersion are essential to reducing exposure, particularly for women.

References:

Shupler, M., Godwin, W., Frostad, J., Gustafson, P., Arku, R. E., & Brauer, M. (2018). Global estimation of exposure to fine particulate matter (PM2. 5) from household air pollution. Environment international, 120, 354-363.

Lee, M., Carter, E., Yan, L., Chan, Q., Elliott, P., Ezzati, M., ... & Baumgartner, J. (2021). Determinants of personal exposure to PM2. 5 and black carbon in Chinese adults: a repeated-measures study in villages using solid fuel energy. Environment International, 146, 106297.

Johnson, M., Lam, N., Brant, S., Gray, C., & Pennise, D. (2011). Modeling indoor air pollution from cookstove emissions in developing countries using a Monte Carlo single-box model. Atmospheric Environment, 45(19), 3237-3243.

Keyword: *Indoor Air Quality, Ventilation, Rural Household*

Healthy Buildings: Scaling Built Environments with Improved Indoor Air Quality

Deepanjali Wadkdar * , **Nidhi Gupta, Gurneet Singh**
Environmental Design Solutions, New Delhi, India

Over the past decade, indoor air quality has garnered significant attention from health professionals, researchers, academia, and policymakers. It has become clear that exposure to contaminated indoor air can have serious adverse health effects. Despite the urgency to address indoor air pollution; building design, construction and management decisions are often made without considering it. This paper is an attempt to identify the barriers and propose an approach to scale healthy built environments using a 5A approach (where each of the A's represent one of the following attributes - Availability, Awareness, Accessibility, Affordability, and Acceptance). The study is an outcome of close consultation with over 100 stakeholders representing owners, building professionals, material and technology providers, facility managers and organisations certifying buildings. Key findings were (a) familiarity of indoor air quality parameters such as carbon dioxide and particulate matter was more than that of formaldehyde and ozone; b) the awareness required for uptake of measures for improvement of indoor air quality was not enough; (c) availability and access to expertise on indoor air quality was expressed as a concern by more than half of the respondents; (d) ease of access and acceptance of rating systems was more of a concern than availability or affordability. The paper surmises that the three levers for market transformation towards built environments with good indoor air quality would be (a) an enabling policy environment backed with mechanisms for implementation; (b) generation of reliable and contextual evidence, (c) concerted efforts on building capacity and creating awareness.

Keyword: *Barrier Assessment, 5A Approach, Market Assessment, Market Transformation*

Promoting Indoor Environmental Quality in Africa: An ISIAQ Innovation Network

Oluyemi Toyinbo * [1], **Kati Huttunen**[2], **Elisephane Irankunda**[3], **Pawel Wargocki**[4], **Adeshokan Muktar**[5], **Ibeh Gabriel Friday**[6], **Jhao-Hong Chen**[7], **Abiyu Kerebo**[8], **Lucmane Koala**[9], **Tunrayo Oluwadare**[10], **Jacob Mensah-Attipoe**[11]

[1] *University of Oulu, Finland*
[2] *Finnish Institute for Health and Welfare, Finland*
[3] *The East African University, Rwanda*
[4] *Technical University of Denmark, Denmark*
[5] *Promoting IEQ/IAQ in Africa ISIAQ Innovation Network*
[6] *Dennis Osadebay University, Asaba Delta State, Nigeria*
[7] *Washington University in St. Louis, St. Louis, USA*
[8] *National Yang Ming Chiao Tung University*
[9] *Université Joseph KI-ZERBO, Ouagadougou, Burkina Faso*
[10] *Afe Babalola University, Ado-Ekiti, Nigeria*
[11] *Savonia University of Applied Sciences, Kuopio, Finland*

Indoor Environmental Quality (IEQ) plays a critical role in shaping our health, comfort, and efficiency, reflecting the reality that modern life keeps us predominantly within enclosed spaces. While extensive IEQ research exists in developed countries, there is a significant gap in the developing world, particularly in Africa. To address this, the International Society of Indoor Air Quality and Climate (ISIAQ) launched the "Promoting IEQ/IAQ in Africa" Innovation Network in 2024, focusing on collaborative research and policy advocacy to improve IEQ across the continent.

The Innovation Network aims to establish a functional African chapter to tackle IEQ challenges unique to the region. By encouraging collaboration among researchers, practitioners, and policymakers, it seeks to promote Africa-specific IEQ research and develop solutions tailored to local needs. Success will be measured by increased regional research output and knowledge-sharing platforms.

In its first year, the "Promoting IEQ/IAQ in Africa" Innovation Network focuses on building inclusive membership, supporting applications for Conference Support Awards, and raising awareness through webinars and editorials. The second year emphasizes quarterly meetings, developing Africa-specific research projects, and sharing findings through ISIAQ webinars. By the third year, the chapter plans to publish a review of African IEQ needs, partner with institutions, and host regional seminars or ISIAQ events in Africa.

The Innovation Network aspires to drive sustainable IEQ advancements, integrating research, innovation, and policy to address Africa's unique challenges and improve public health outcomes.

Keyword: *Africa, Research Collaboration, ISIAQ*

Indoor Environmental Quality Challenges and Student Comfort in East African Classrooms

Oluyemi Toyinbo * [1], **Eunice Jengo**[2], **Xuzel Villavicencio Peralta**[3], **Björn Haßler**[3]

[1] *University of Oulu*
[2] *College of Business Education, Dar es Salaam, Tanzania*
[3] *Open Development and Education, United Kingdom*

Introduction
This study investigates Indoor Environmental Quality (IEQ) in East African secondary school classrooms and its impact on student comfort. Key IEQ factors include temperature, humidity, air quality, noise, and lighting. Conducted under the Improving Learning Through Classroom Experience (ILCE) program, this research addresses the lack of data on IEQ in schools in developing countries.

Method
Fifteen classrooms across five schools in Dar es Salaam, Tanzania, were assessed using walkthrough surveys and subjective comfort assessments from 1,000 students. Observational data captured conditions such as overcrowding, ventilation, and clutter.

Results
Only 1.4% of students reported comfort across all IEQ parameters, while 3.8% were comfortable with four and 13.7% with three. Key issues included excessive noise levels (69%), a desire for fresher air (86%), and overcrowding, with student densities far exceeding recommended levels (35 individuals per 100 m^2) in all classrooms studied. Overcrowding contributed to excessive clutter in all classrooms, impeding effective classroom management. While some students expressed dissatisfaction with temperature, 76% preferred the current conditions, reflecting adaptive thermal comfort preferences.

Conclusion
The study highlights the need for region-specific interventions using locally sourced, cost-effective materials. Proposed solutions include white roof coatings, shading devices, and enhanced ventilation. These findings contribute to understanding IEQ in developing countries and advocate for sustainable, equitable learning environments. Ongoing evaluations aim to refine interventions and improve classroom conditions in East Africa.

Keyword: *East Africa, Indoor Environmental Quality, Student Comfort*

Assessment of Energy and Daylight Models Across Building Typologies for High-Performance Design

Deepashree KS * [1], **Nikita Khatri**[2]
[1] *CEPT University, Ahmedabad, India*
[2] *Ecofirst Services Limited, Bengaluru, India*

The intersection of energy performance and natural daylighting in buildings is crucial for achieving high-performance structures. Energy and daylight modelling are essential tools for analysing how building functions influence energy use and daylight availability, enabling the optimization of energy and occupant comfort.

This research aims to analyse the impact of building functions on energy demand and daylight distribution, examining how occupancy patterns, spatial layouts, and design choices affect performance. By leveraging simulation tools, the study intends to provide practical design recommendations for optimizing window placement, shading strategies, interior layouts, and HVAC systems to reduce energy consumption and enhance thermal comfort in buildings.

As cities and buildings grow increasingly complex, with diverse functional requirements, the need for such assessments is becoming more critical. Exploring energy and daylight modelling through the lens of building function offers valuable insights into how different spaces interact with natural and artificial energy systems.

Furthermore, establishing a correlation between energy use intensity (EUI) ranges of various building typologies with their respective climate types can help formulate standard practices to ensure measurable improvements in overall building performance.

By addressing these challenges, this research contributes to the development of smarter building designs that are not only energy-efficient but also promote healthy, comfortable living environments for the future.

Keyword: *High-performance buildings, energy & daylight simulations, Building typologies*

Development of Residential Air Conditioning Equipment for Health Risk Prevention and Health Promotion -Subject experiment on the effects of different room temperature lowering speeds on physiological responses and psychological declarations (in winter)

Sakyo Okada *
Chiba University, Chiba, Japan

"Extending healthy life expectancy" is attracting attention as a solution to the increasing social insurance costs associated with the aging of society.

To realize this theme, this study focused on the air conditioning systems of residences. Current air conditioning systems have the risk of degrading the adaptive functions of the human body and are insufficient in terms of consideration for the health of residents. Therefore, a control system that fluctuates the room temperature within a range that does not cause discomfort to the human body is considered to provide a comfortable thermal environment while giving an appropriate load to the autonomic nervous system.

In this paper, an experiment was conducted in which subjects experienced two different speeds of room temperature decrease from the same room temperature during the winter season. The purpose of the experiment is to compare the differences in psychological reports of warmth, coldness, and comfort, and the differences in various physiological indices, and to provide a foothold to consider appropriate room temperatures and room temperature change speeds.

The results of the experiment showed that physiological indices such as deep body temperature and skin surface temperature fluctuated in those with a slower rate of descent, even within the same room temperature change range. No difference in room temperatures perceived as cold or uncomfortable was observed for the different descent speeds. This indicated that a gradual change in room temperature may moderately stimulate both the surface and deep physiological mechanisms of the human body.

Keyword: *Indoor thermal environment, room temperature change rate, human body's adaptive function*

Surviving indoor heat stress in Indian cities: Urban to building-scale impacts assessment on occupant thermal comfort

Nilabhra Mondal * , **Prashant Anand**
Indian Institute of Technology Kharagpur

Climate, technologies, and socio-economic acclimatization collectively shape perceptions of indoor thermal comfort, significantly influencing the future use of building cooling systems in urban areas. With the intensification of extreme climates, urban heat stress has emerged as a critical public health challenge, necessitating urgent measures to mitigate building overheating and ensure adequate cooling. In Indian cities, where air conditioning penetration remains limited, mixed-mode ventilation and cooling systems face increasing challenges in addressing summer heat stress. Current metrics, such as heat-health warning systems, inadequately assess human-centric indoor heat stress and the performance of HVAC systems in ensuring comfort. This study develops a top-down observational and simulation-based framework to evaluate heat stress impacts on human thermal comfort and building-to-urban scale cooling energy demand, focusing on Indian cities during the summer months. The framework estimates indoor cooling loads based on regional climate, urban morphology, and microclimatic conditions. It also dynamically links outdoor temperature fluctuations with indoor heat exposure, accounting for building archetypes and occupant thermal perceptions. The study seeks to enhance indoor adaptive comfort models in the Indian context by identifying possible set-point variability for mixed-mode HVAC systems during summer months and offering actionable control strategies. Furthermore, the framework evaluates whether the increasing penetration of air conditioning across urban areas and associated indoor controls can effectively mitigate heat stress. The outcomes aim to provide generalized insights applicable to developing regions in South Asia, Africa, and the Middle East, addressing spatial variability in heat stress exposures within urban patches of varying densities and urbanization levels.

References:

Hampo, C. C., Schinasi, L. H., & Hoque, S. (2024). Surviving indoor heat stress in United States: A comprehensive review exploring the impact of overheating on the thermal comfort, health, and social economic factors of occupants. Heliyon.

Laurent, J. G. C., Williams, A., Oulhote, Y., Zanobetti, A., Allen, J. G., & Spengler, J. D. (2018). Reduced cognitive function during a heat wave among residents of non-air-conditioned buildings: An observational study of young adults in the summer of 2016. PLoS medicine, 15(7), e1002605.

Yi, C. Y., Peng, C., & Robinson, D. (2023). Modelling urban dwellers' indoor heat stress to enhance heat-health warning and planning. Building and Environment, 243, 110623.

Keyword: *Indoor overheating, Indoor heat stress, Thermal comfort*

A Deep reinforcement learning agent for Optimizing Personal Exposure, Energy consumption, and thermal Comfort (DOPEEC)

Nishchaya Kumar Mishra * , **Nipun Batra, Sameer Patel**
Indian Institute of Technology Gandhinagar, India

The indoor environments of buildings directly impact occupants' well-being. Since people spend almost 90% of their time indoors, well-controlled indoor environments are imperative to reduce pollutant exposure and ensure thermal comfort. However, a trade-off exists between energy consumption, exposure reduction, and thermal comfort, necessitating optimization. While fully physics-driven strategies have been widely studied for the same purpose, their large-scale integration with buildings is challenging due to computational and modeling limitations. Reinforcement learning (RL) agents that overcome the limitations of physics-driven strategies have the potential to learn and control different indoor environments. This work developed a deep RL agent (DOPEEC: Deep reinforcement learning agent for Optimizing Personal Exposure, Energy, and Comfort) to optimize exposure, energy, and thermal comfort in a house by controlling the ventilation rate and the set temperature. Agent 1 dynamically changed the ventilation rate, and Agent 2 controlled the ventilation rate and the indoor set temperature. A sensitivity analysis for different combinations of weights in the reward function has been conducted, and the performance of DOPEEC is compared to a physics-based dynamic optimization (DynOpt) strategy. Compared to 'only ventilation control' at a fixed set temperature of 25°C, the dynamic set temperature control resulted in comparable exposure reduction but with lower energy penalties, i.e., higher NERs. This work demonstrates the potential of RL agents for widely deployable solutions to control building environments to safeguard occupants' health while minimizing environmental impacts.

Keyword: *deep reinforcement learning, energy-exposure-comfort trade-off, indoor air pollution control, dynamic optimization, variable indoor-outdoor air exchange rate, dynamic set temperature*

Realtime Optimization of Personal Exposure to Particulate Matter, Energy Consumption, and Thermal Comfort in a House

Nishchaya Kumar Mishra * [1], **Marina E. Vance**[2], **Atila Novoselac**[3], **Sameer Patel**[1]
[1] *Indian Institute of Technology Gandhinagar, India*
[2] *University of Colorado, Boulder, USA*
[3] *The University of Texas at Austin, USA*

People worldwide spend more than 80% of their time in the built environment, and thus, it becomes imperative to assess indoor environment quality (IEQ), which affects human health, comfort, and productivity. IEQ is primarily controlled by indoor air quality (IAQ) and heating, ventilation, and air conditioning (HVAC) systems. The HVAC operation in buildings accounts for a significant fraction of global energy consumption. Measures to improve IAQ can further increase the energy demand. Thus, optimizing IAQ, energy consumption, and thermal comfort is critical.

This work presents a real-time optimization model to investigate the complex and interdependent relationship between personal exposure to particulate matter (PM), thermal comfort, and energy consumption in a house. Multiple scenarios for regular and intensive PM emission in the house, along with different HVAC configurations and ambient conditions, have been investigated. An aerosol dynamic model integrated with a physics-based model of HVAC operation is developed. The HVAC system has been modeled to maintain the house's thermal comfort and air recirculation. Further, an objective function is defined to optimize the model and penalize the exceedance of thresholds for indoor pollutant concentration and indoor temperature. The optimization strategy minimized the objective function constituting energy consumption, exposure, and thermal comfort. This work also discusses the challenges associated with the large-scale deployment of similar optimization strategies. A combination of filtration, exfiltration, and dynamic indoor temperature is crucial to minimize the energy penalty associated with exposure reduction while ensuring occupants' thermal comfort.

References:

Mishra, N. K., M. E. Vance, A. Novoselac, and S. Patel. 2024. "Dynamic optimization of personal exposure and energy consumption while ensuring thermal comfort in a test house." Build. Environ. 252 (Jun): 111265. https://doi.org/10.1016/j.buildenv.2024.111265.

Keyword: *dynamic optimization, energy-exposure tradeoff, indoor air pollution mitigation, thermal comfort, healthy indoor environment*

Assessing the Impact of Indoor Environmental Conditions, $PM_{2.5}$ Levels, on Creativity Using EEG and Sensory Devices

Farhan Asim[1], Shreya Rai * [1], P S Chani[2], Venu Shree[3], Sandeep Budde[4]

[1] *Built Environment Lab, Department of Architecture and Planning, Indian Institute of Technology, Roorkee - 247667, India*

[2] *Built Environment Laboratory, Department of Architecture & Planning, IIT Roorkee, Roorkee 247667, India*

[3] *Department of Architecture, NIT Hamirpur, India*

[4] *Urban Environment Observatory, School of Urban and Regional Planning, University of Alberta, Edmonton AB - T6G 2R3, Canada*

Indoor environmental conditions, including air quality and thermal parameters, significantly influence cognitive functions, yet their impact on creativity remains underexplored. This study investigates the associations between indoor air quality particularly particulate matter ($PM_{2.5}$), carbon dioxide (CO_2) levels, and thermal conditions and creativity of different group sizes within a 12 people capacity lab, integrating behavioral and neurophysiological analyses. Conducted in a controlled lab environment at IIT Roorkee, the research employs (EEG) and sensory devices to monitor brain activity and physiological responses, while environmental sensors measure $PM_{2.5}$ concentrations, CO_2 levels, temperature, and humidity. Creativity is assessed using validated tools such as the Alternative Uses Test, with EEG providing insights into neural correlates of cognitive flexibility, fluency, and originality. This study hypothesizes that lower $PM_{2.5}$ and CO_2 levels, combined with normative thermal conditions (22-26°C), enhance creative thinking and optimize neural activity patterns. By linking environmental parameters with EEG-derived biomarkers, the research aims to uncover the neural mechanisms influenced by indoor air quality and thermal conditions. Findings will inform strategies for designing indoor environments that support creativity and cognitive performance, benefiting workplaces, educational settings, and residential spaces.

References:

Asim, F., Chani, P. S., & Shree, V. (2021). Impact of COVID-19 containment zone built-environments on students' mental health and their coping mechanisms. Building and Environment, 203(July), 108107. https://doi.org/10.1016/j.buildenv.2021.108107

Asim, F., Chani, P. S., Shree, V., & Rai, S. (2023). Restoring the mind: A neuropsychological investigation of university campus built environment aspects for student well-being. Building and Environment, 244(August), 110810. https://doi.org/10.1016/j.buildenv.2023.110810

Dedesko, S., Pendleton, J., Petrov, J., Coull, B. A., Spengler, J. D., & Allen, J. G. (2025). ASSOCIATIONS BETWEEN INDOOR ENVIRONMENTAL CONDITIONS AND DIVERGENT CREATIVE THINKING SCORES IN THE COGFX GLOBAL BUILDINGS STUDY. Building and Environment, 112531. https://doi.org/10.1016/j.buildenv.2025.112531

Keyword: *Indoor air quality, $PM_{2.5}$ levels, Electroencephalography (EEG)*

H&W Certification Tool for Enhancing Occupational Health and Proposed Interventions in Building Design & Operation

Shivraj Dhaka * [1], **Nisha Dugar**[2]
[1] *Senior Counselor - CII-Indian Green Building Council*
[2] *Architect - CII-Indian Green Building Council*

Study of work environment is crucial and it impacts Health & wellbeing of people. Studies indicate out of 100% of the building's operational costs, 90% accounts for human resource costs. It increases owing to poor IEQ. There are no tools available to implement for improvement in work environment. IGBC developed H&W tool to resolve the issue. The tool used to certify 20 office Buildings in 7 major cities, India. To further assist the certification, research carried out to understand the Impact of Green Buildings on H&W. Primarily, through the tool, building design and operational characterises collated and examined based on several criterion developed for 7 key areas those affects work environment. To study various thresholds defined by the ISHRAE IEQ 10001 the physical measurements and occupant satisfaction through surveys on the IEQ was collected from projects. The results indicate interventions in the building design and also made recommendations for the existing spaces, to meet and exceed performance parameters prescribed in the H&W certification. It has noted that more than 70% occupants were satisfied within a temperature range of 26$\pm$2 degC, RH 30%-70% and air velocity of $<$0.2m/s, this offered to save more energy and supported occupants to feel thermally more comfortable. The tool enabled designers to explore and integrate measures with the new design, to meet occupant comfort and it has recommended measures to alter work environment to achieve H&W certification. The study concluded the impact of active, passive measures on thermal, visual, air quality, acoustic and ergonomics.

References:

https://igbc.in/frontend-assets/html_pdfs/DI%20-%20IGBC%20Sustainable%20Workspaces%20and%20Indian%20Business%20Environement.pdf
https://igbc.in/frontend-assets/html_pdfs/Healthy%20Buildings%20-%20A%20Need%20of%20the%20Hour.pdf
https://imsear.searo.who.int/items/223ef836-fbdd-4ddb-ac99-1ff955047f88

Keyword: *Indoor Environmental Quality, Indoor Air Quality, Occupant Well-Being*

ENERGY AND CAPEX ANALYSIS FOR AIR PURIFICATION AND VENTILATION STRATEGIES IN RESIDENTIAL BUILDINGS

Shubham Solanki * , **Anmol Mathur**
COLEAD LLP, New Delhi, India

ABSTRACT

This study examines optimal strategies for improving indoor air quality in residential buildings by analyzing the energy and capital expenditure (CAPEX) implications of centralized versus decentralized air purification systems. The study also evaluates the impact of outdoor airflow rates on CO2 concentration control and associated energy consumption.

Centralized systems, incorporating MERV 13 and HEPA filters, demonstrated uniform air quality improvement and moderate operational costs, whereas decentralized systems offered localized control at the expense of higher CAPEX and energy usage. For CO2 control, maintaining an outdoor airflow rate of 2.5 L/s per person, as per ASHRAE 62.1 standards, was sufficient to keep CO2 concentrations below 1000 ppm. Doubling this airflow reduced CO2 levels to 500 ppm, significantly enhancing occupant comfort and health.

Increased outdoor airflow rates, while improving air quality, led to a notable rise in HVAC energy consumption. Quadrupling the standard airflow resulted in a 26.4% increase in total energy use. Demand-controlled ventilation systems, utilizing real-time occupancy and CO2 sensors, were recommended to optimize energy efficiency without compromising air quality.

This research offers actionable insights into designing energy-efficient, cost-effective air purification and ventilation systems, balancing sustainability with occupant health and comfort.

Corresponding email: info@colead.in

Keyword: *Indoor air quality, air purification systems, CO2 control, HVAC energy, ventilation optimization*

Assessing the Influence of Different Ventilation Strategies on Indoor Radon Levels: A Case-Control Study.

James McGrath *
Maynooth University

Radon is a radioactive gas and is a recognised leading cause of lung cancer in most countries. Radon infiltrates into buildings from the trace amount of uranium in soil, where it can accumulate to harmful levels. However, the rate of infiltration varies depending on environmental conditions within and outside the home. As Near Zero Energy Buildings construction becomes increasingly prevalent, understanding the interplay between ventilation systems and radon mitigation is vital.

This study investigates the impact of different ventilation strategies on indoor radon levels in newly constructed dwellings. The research utilises ten identical unoccupied, newly constructed, energy-efficient Irish dwellings to minimise confounding variables and ensure statistical relevance. The statistical experimental design is based on three ventilation treatments: no mechanical ventilation, mechanical ventilation with heat recovery, and simulated continuous mechanical extract ventilation. Digital detectors will monitor hourly radon levels in two rooms per dwelling. To ensure statistical relevance, the dwellings will be divided into three groups, and each group will receive a matrix combination of testing weeks.

The study aims to evaluate the effectiveness of each ventilation strategy in reducing indoor radon levels. Statistical models will analyse the data, considering local weather conditions recorded by local weather stations. This research will provide valuable insights into the effectiveness of various ventilation strategies in managing indoor radon levels in modern, energy-efficient homes.

Keyword: *Radon, Ventilation, Airtight*

Effect of natural and mechanical ventilation in anticipating the occurrence of sick building syndrome (Case study: Pharmaceutical Laboratory in Surabaya)

Aprisia Rasya Murran *
University of Brawijaya

Ventilation in laboratory is crucial for diluting chemical hazards and ensuring air exchange. Poor indoor air quality (IAQ) can lead to Sick Building Syndrome (SBS), often linked to inadequate ventilation, indoor pollutants, and occupant behavior. Three chemical laboratories were examined: wet lab, a dry lab, and an acid room that simultaneously utilized air conditioning (AC) and operable windows, contrary to standard guidelines. This study aimed to: (1) assess physical IAQ conditions in laboratories, (2) evaluate the effects of natural and mechanical ventilation modes, and (3) explore how IAQ impacts SBS complaints.

A mixed-method approach was employed, combining qualitative ventilation system assessments based on design standards with quantitative IAQ monitoring of temperature, humidity, air velocity, and air changes per hour (ACH). Four scenarios were tested: AC on/off with windows open/closed. Results were compared to national guidelines, and prior SBS complaints were reviewed.

All laboratories met the guideline of 6 ACH. Natural ventilation with open windows provided significantly higher ACH values (9.42–15.85) due to increased air velocity, while mechanical systems achieved controlled but lower ACH ranges (6.33–8.68). However, the design of window sizes and AC capacities was insufficient, and a temperature rise of 0.5–0.6°C was observed with AC use, influenced by solar radiation and window dimensions.

Both ventilation modes reduced SBS risks, but natural ventilation was more effective due to enhanced air circulation from outdoor winds. Design improvements, particularly in window sizing and AC capacity, are recommended to optimize overall ventilation performance.

References:

ASHRAE. (2018). Classification of Laboratory Ventilation Design Levels. Atlanta: ASHRAE. Guan, L. (2011, September). The Influence of Glass Types on the Performance of Air-Conditioned Office Buildings in Australia. Advanced Materials Research, vol. 34, pp. 34-39.
Ratodi, M. (2017). Predicting the Sick Building Syndrome (SBS) occurrence among Pharmacist assistant in Banjarmasin South Kalimantan. Health Science Journal of Indonesia, 118-123.
Yau et al. (2012). Studies on the Indoor Air Quality of Pharmaceutical Laboratories in Malaysia. International Journal of Sustainable Built Environment, 110-124.

Keyword: *Indoor air quality, ventilation, pharmaceutical laboratory*

Integrated Measurement and Simulation for Optimizing Indoor Air Quality in Biosafety Laboratories

Nitesh Singh * [1], **Hamza H. S. Helal[1], Rahmai Aulia Pramudantoro[2], Fu-Jen Wang[2]**
[1] *Graduate Institute of Precision Manufacturing, National Chin-Yi University of Technology, Taichung 41170, Taiwan*
[2] *Department of Refrigeration, Air Conditioning and Energy Engineering, National Chin-Yi University of Technology, Taichung 41170, Taiwan*

Ensuring optimal indoor air quality (IAQ) in critical environments, such as hospital biosafety laboratories (BSL), is vital for safety and operational efficiency. This study introduces a replicable methodology for assessing and optimizing IAQ through a combination of field measurements and Computational Fluid Dynamics (CFD) simulations. The methodology was applied at the Taichung Veterans General Hospital Biosafety Laboratory to evaluate key parameters, including air volume, pressure gradients, airflow patterns, temperature, and humidity levels. Results revealed compliance with design standards, such as air exchange rates of 12 ACH ±5%, pressure gradients exceeding 12.5 Pa, and stable indoor conditions of 21 ±2°C and 60 ±10% RH. Airflow smoke tests confirmed unidirectional flow, mitigating cross-contamination risks. Illuminance levels (500–760 lux) and noise levels (<60 dBA) met or surpassed operational criteria, ensuring a conducive workspace. However, CFD simulations highlighted insufficient airflow and uneven distribution in Room 2 (P2-P3 lab), an issue corroborated by field measurements. Four design configurations were simulated to address this, revealing that Design 3 offered the best airflow uniformity and thermal distribution, minimizing stagnation and hot spots. This integrated approach demonstrates the effectiveness of CFD as a diagnostic tool and provides a replicable framework for enhancing IAQ and energy efficiency in healthcare and lab facilities.

Keyword: *Biosafety Laboratory, CFD Simulation, Indoor Air Quality*

Indoor Environmental Quality-Ventilation-Energy Nexus: Rethinking Indoor Environments

Khyati Upadhyay, Manoranjan Sahu *
Indian Institute of Technology Bombay

Indoor environmental quality (IEQ) plays a crucial role in ensuring the health, comfort, and productivity of occupants, particularly in educational institutions where prolonged exposure to indoor air pollutants and suboptimal thermal conditions can significantly impact well-being and performance of students. This study focuses on assessing the IEQ of the Makerspace Lab at IIT Bombay, a high occupancy space with a special emphasis on the relationship between indoor air pollution, thermal comfort, and energy consumption. The study investigates key IEQ parameters, including $PM_{2.5}$, PM_{10}, CO_2, humidity, and temperature under different laboratory conditions. The findings show a distinct correlation between lab activities and pollutant concentrations, with machine demonstrations contributing to elevated levels of $PM_{2.5}$ well above the acceptable range of 15 $\mu g/m^3$. CO_2 concentrations consistently exceeded the recommended threshold of 1000 ppm. Maximum CO_2 concentrations observed varied from 2000 to 3500 ppm and remained very high during operational hours, underscoring the need for enhanced ventilation strategies. Relative humidity measurement indicates 15-20% time the value exceed the recommended value of 70%. Results indicate that certain lab activities such as soldering, machining in lazer cutting machine lead to spikes in pollutant levels. The persistent high CO_2 levels suggest that the current ventilation system is insufficient in maintaining optimal air quality during periods of high occupancy. In addition to evaluating existing conditions, the detailed results of the optimized HVAC system operation, focusing on identifying a flow rate that maximizes indoor air quality (IAQ) and thermal comfort will be presented.

Keyword: *Indoor Environmental Quality, Indoor air quality, thermal comfort*

AI-Driven Building Management: Bridging Gaps in Automation to Enhance Indoor Environmental Quality and Energy Efficiency.

Arsh Arora * , **Kevin Joshi, Ronak Sutaria**
Respirer Living Sciences, Pune , India

The global drive for healthier and energy-efficient buildings has highlighted the need for advanced automation technologies. While AI-driven systems show promise in managing indoor environmental quality (IEQ), significant gaps persist due to insufficient data, limited contextual adaptability, and scalability challenges.

This paper introduces an AI-driven building management system designed to address these issues. The system uses a comprehensive network of IOT based sensor network to monitor real-time data on air quality, thermal comfort, and estimate energy usage across building zones. AI algorithms analyse this data to provide actionable insights and tailored recommendations, such as optimizing HVAC settings and employing passive ventilation strategies. These insights are communicated effectively to operating systems, facility managers, and occupants, creating a dynamic feedback loop that enhances decision-making and automation.

The approach addresses key gaps through:

Comprehensive Data Collection: Capturing critical parameters to provide a holistic understanding of the environment. **Contextual Adaptability:** AI evolves with the unique operational needs of each building. **Scalability:** Modular hardware and cloud integration enable application across buildings of varying sizes.

Field implementation in an IGBC Platinum-rated corporate office demonstrated significant results: a 47% improvement in IAQ compliance, consistent CO_2 levels under 800 ppm, and energy savings of 18.5% through demand-controlled ventilation and passive cooling.

This paper underscores the transformative potential of AI-driven solutions in achieving healthier, more sustainable buildings and provides a roadmap for overcoming current limitations in automation technologies

Keyword: *indoor air quality, internet of things, sick building syndrome*

Assessment of Indoor Air Quality across microenvironments: Implications for Particulate Matter and CO_2 exposure management

Riya Verma *, **B.R. Gurjar, C.S.P. Ojha**
Indian Institute of Technology, Roorkee, India

The health of occupants and their work performance highly depend on the indoor environment in which they work. This case study focuses on assessing the IAQ within different microenvironments. Three different types of microenvironments (residence, office, and school classroom) were considered for this analysis. The process of analysis involves the evaluation of coarse, fine, and submicron particles (PM_{10}, $PM_{2.5}$, and PM_1) and CO_2 levels inside the considered microenvironments. The hourly average concentrations for PM_{10}, $PM_{2.5}$ and PM_1 were found to be (148.91 ± 169.76), (75.45 ± 32.11) and (58.06 ± 25.49) for residence; (62.49 ± 42.62), (34.22 ± 14.70) and (23.49 ± 8.75) for office; (118.44 ± 65.07), (50.49 ± 17.61) and (32.53 ± 11.60) for classroom respectively. The CO_2 levels were (667.91 ± 103.86) for residence, (816.87 ± 339.71) for office, and (714.67 ± 348.08) for classroom respectively. For the residential environment and the classroom, the critical pollutant of concern was particulate matter; for office spaces, the critical pollutant of concern was CO_2. Fractional analysis of particulate matter for different sizes provides critical insights into the sources, behavior, and health impacts of particulate pollution. The results of the current study will find implications in exposure assessment and help the regulating authorities to develop a policy framework for Indoor Air Quality management.

Keyword: *Indoor Air Quality (IAQ), Particulate Matter (PM), CO2 Levels, microenvironments*

Assessment of Indoor Air Quality Across Microenvironments: Implications for Particulate Matter and CO_2 Exposure Management

Riya Verma[1], B.R. Gurjar[1]*, C.S.P. Ojha[1]
[1]Department of Civil Engineering, Indian Institute of Technology, Roorkee, India
*Corresponding email: bhola.gurjar@ce.iitr.ac.in

SUMMARY

The health of occupants and their work performance highly depend on the indoor environment in which they work. This case study focuses on assessing the IAQ within different microenvironments. Three different types of microenvironments (residence, office, and school classroom) were considered for this analysis. The process of analysis involves the evaluation of coarse, fine, and submicron particles (PM_{10}, $PM_{2.5}$, and PM_1) and CO_2 levels inside the considered microenvironments. The hourly average concentrations for PM_{10}, $PM_{2.5}$ and PM_1 were found to be (127.1 ± 54.7), (74.3 ± 27.9) and (57.1 ± 22.1) for residence; (62.5 ± 42.6), (34.2 ± 14.7) and (23.5 ± 8.7) for office; (118.4 ± 115.2), (50.5 ± 17.6) and (32.5 ± 11.6) for classroom respectively. The CO_2 levels were (667.9 ± 90.8) for residence, (816.8 ± 339.7) for office, and (714.6 ± 348.1) for classroom respectively. For the residential environment and the classroom, the critical pollutant of concern was particulate matter; for office spaces, the critical pollutant of concern was CO_2. Fractional analysis of particulate matter for different sizes provides critical insights into the sources, behaviour, and health impacts of particulate pollution. The results of the current study will find implications in exposure assessment and help the regulating authorities to develop a policy framework for Indoor Air Quality management.

KEYWORDS

Indoor Air Quality (IAQ); Particulate Matter (PM); CO_2 Levels; microenvironments

1 INTRODUCTION

In recent times, concerns regarding Indoor Air Quality (IAQ) have emerged globally, mainly because individuals now spend a significant portion of their time within different indoor microenvironments such as offices, residences, fitness and wellness centres, educational institutions, and other commercial spaces like shopping complexes, etc[1]. These indoor microenvironments can lead to exposure to a wide range of indoor contaminants such as particulate matter (PM_{10} and $PM_{2.5}$), gaseous emissions such as volatile organic compounds (VOCs), oxides of carbon, ozone (O_3) etc, and other biological contaminants such as bacteria, fungi, etc [2]. The primary sources of these indoor contaminants include building and furnishing materials, cleaning agents, air fresheners, and everyday human activities such as cooking or cleaning, etc. Compounding the issue, inadequate ventilation and air tightness of buildings can trap these pollutants and hinder their effective dispersion thereby worsening the air quality. Infiltration of pollutants from outdoors through gaps or vents may also increase the level of indoor pollution [3].

Since the quality of indoor air is influenced by the functions and characteristics of each microenvironment, it is essential to identify the potential sources, exposure pathways, and dispersion patterns for each microenvironment. Therefore, the present study focuses on understanding the temporal and spatial variations of indoor pollutants within the considered microenvironments.

2 METHODS

Site description

For the present study, 3 different sites (residence, office and classroom) were selected for monitoring of indoor air quality. These sites were in Delhi-NCR region. The residence was a 2 BHK flat located in Noida having total area of 1500 sqft. It had a natural ventilation system with air-conditioners running during afternoon and night. Total number of occupants living in the house were 8 which included kids and elderly. The office space considered for the study was in Gurugram, Haryana and had a total area of approximately 2500 sqft. It was equipped with mechanical ventilation system and was functional from 9:00 am to 7:00 pm. Around 15-20 people were working in the considered office space. The school classroom considered for the study was also located in Gurugram. It was a classroom in primary wing with a total area of 350 sqft and was equipped with mechanical ventilation system. Total occupancy for the class including the kids and the staff was around 30-35.

Monitoring

IAQ monitoring for particulate matter (PM_{10}, $PM_{2.5}$, and PM_1) and CO_2 was performed at the selected microenvironments for 24 hours. Samples were collected for 5 days at a sampling rate of 1 min using environmental dust monitor (GRIMM make, Model 1.109). The 31-channel dust monitor employed in this study operates on the principle of light scattering. It samples air at a flow rate of 1.2 L/min and detects airborne particles within the size range of 0.3 to 25 μm. Particle number concentrations are subsequently converted to mass concentrations using established calibration algorithms[4]. The instrument was positioned at a height of approximately 1.3 m above the floor and at least 1 m away from any walls. The inlet head was placed as close as possible to the typical breathing zone of the occupants to ensure representative exposure measurements. CO_2 levels were measured using Gray Wolf's portable multi gas monitor (TG-503L, Gray Wolf Sensing Solution, Shelton, USA).

3 RESULTS

Temporal variation of parameters

The hourly average concentrations of pollutants were found to vary with the working hours within the considered microenvironments. Figure ***1*** describes the temporal variation of parameters within studied indoor environments. PM concentrations in the residence exhibited two different peaks, one in the morning (08:00– 11:00) and another in the evening (08:00–10:00), most likely as a result of daily tasks like cleaning or cooking. Following this, CO_2 levels rose after 7:00 am and peaked at 12:00 pm (~850 ppm), continuing to be high throughout the evening.

In the office setting, PM levels were comparatively low in the morning but increased significantly between 10:00 am to 12:00 pm and 06:00 pm to 08:00 pm in the evening, as the movement was maximum at this time. Similar trends were seen in CO_2 concentrations, which peaked at about 1500 ppm at midday, suggesting excessive occupancy and inadequate ventilation[5].

The classroom exhibited the most pronounced pattern, with both PM and CO_2 concentrations increasing significantly between

10:00 am and 02:00 pm (PM_{10} peaking at ~450 µg/m³ and CO_2 at ~1400 ppm, coinciding with classroom occupancy. After 03:00 pm, both parameters returned to baseline levels.

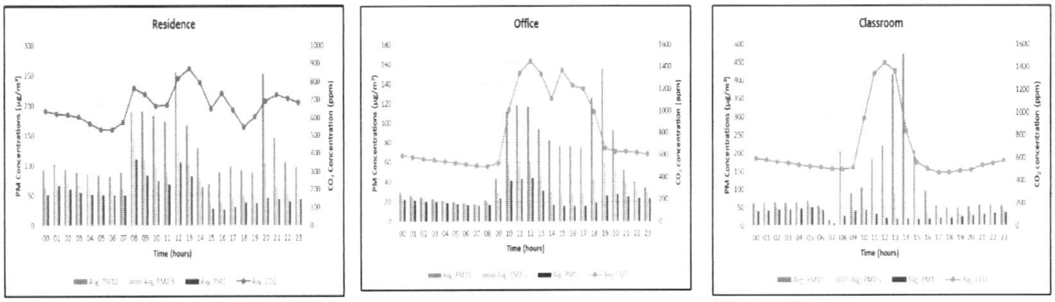

Figure 1:Temporal variation of parameters at residence, office, and classroom.

Spatial variation of parameters

Spatio-temporal assessment has been carried out for each microenvironment and the descriptive statistics for the considered indoor pollutants are given in Table 1. The residence had the highest daily average concentrations of PM_{10} (127.1 ± 54.7 µg/m³) and $PM_{2.5}$ (74.3 ± 27.9 µg/m³) among the sites, indicating increased particle-generating activities like cooking or cleaning. The classroom showed the maximum PM_{10} value (472.6 µg/m³) and higher PM levels overall, most likely as a result of high occupancy and activities of children during school hours. On the other hand, the office had reduced PM concentrations in all size fractions, suggesting a more regulated setting with fewer sources of particulate matter (PM_{10}: 62.5 ± 42.6 µg/m³ and $PM_{2.5}$: 34.2 ± 14.7 µg/m³). The office had the highest average concentration of CO_2 (816.8 ± 339.7 ppm), which may indicate inadequate ventilation and prolonged occupancy. The high standard deviation of CO_2 levels in classrooms (±348.1 ppm) and offices (±339.7 ppm), compared to residences (±90.8 ppm), suggests greater variability in ventilation and occupancy patterns in these spaces, likely linked to scheduled use during working or school hours. These findings highlight how ventilation, occupancy patterns, and human activity affect indoor air quality in various microenvironments.

Table 1: Descriptive statistics for indoor pollutants in studied environments.

Parameter	Sites →	Residence	Office	Classroom
PM_{10} (ug/m³)	Min.	69.6	16.9	15.1
	Max.	257.1	189.4	472.6
	Mean± SD	127.1 ±54.7	62.5 ±42.6	118.4 ± 115.2
$PM_{2.5}$ (ug/m³)	Min.	39.5	16.6	8.9
	Max.	147.1	67.5	98.3
	Mean± SD	74.3 ±27.9	34.2 ± 14.7	50.5 ± 17.6
PM_1 (ug/m³)	Min.	27.1	14.3	7.5
	Max.	110.7	47.1	50.9
	Mean± SD	57.1 ±22.1	23.5 ± 8.7	32.5 ± 11.6
CO_2 (ppm)	Min.	532.5	485	467.6
	Max.	874.2	1558	1436.6
	Mean± SD	667.9 ±90.8	816.8 ±339.7	714.6 ± 348.1

Indoor Air Quality status

By comparing pollutant concentrations with the suggested recommendations and health standards for indoor pollutants, one can assess the IAQ status of a microenvironment. Since, no such standards or guidelines are available for India, we have compared the 24-hour average pollutant concentration to American Society of Heating, Refrigerating and Air-Conditioning Engineering (ASHRAE), and World Health Organization (WHO) air quality guidelines.

The 24-hr average concentrations of PM_{10}, $PM_{2.5}$, and CO_2 against corresponding recommended values are shown in Figure 2. The recommended values marked by red lines reflect WHO guideline values for PM_{10} (50 µg/m³) and $PM_{2.5}$ (25 µg/m³), and the ASHRAE recommended limit for CO_2 (1000 ppm). The data reveal that daily average PM_{10} concentrations in both residence (127.1 µg/m³) and classroom (118.4 µg/m³) exceed the WHO limit to a much greater extent as compared to office levels (62.5 µg/m³) indicating significant sources of particulate matter in residence and classroom. A similar pattern is observed for $PM_{2.5}$, where daily average values in residence (74.3 µg/m³) and classroom (50.5 µg/m³) are well above the WHO threshold, and office levels (34.2 µg/m³) remain moderately elevated. In contrast, the 24-hr average CO_2 concentrations across all microenvironments (residence: 667.9 ppm, office: 816.8 ppm, classroom: 714.6 ppm) fall below the ASHRAE guideline, suggesting generally adequate ventilation. Overall, the results highlight poor indoor air quality with respect to particulate matter, particularly in residences and classroom, while CO_2 levels remain within acceptable limits across all environments.

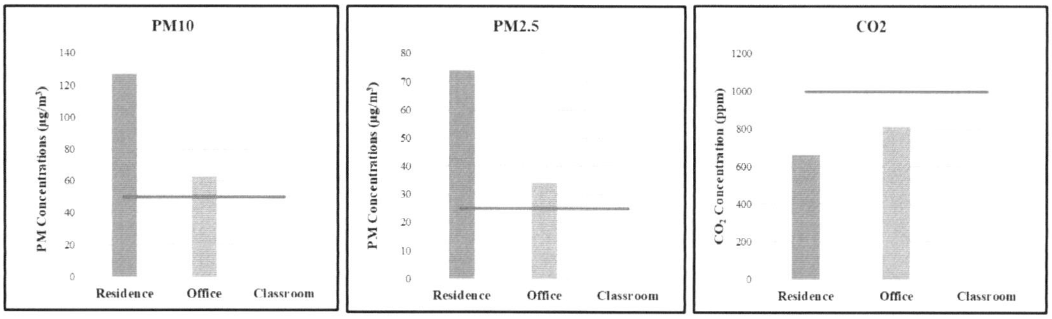

Figure 2: Daily average concentration of PM_{10}, $PM_{2.5}$ and CO_2 in different Indoor Microenvironments. Corresponding recommended values shown via horizontal red lines are: WHO guidelines values for PM_{10} and $PM_{2.5}$ and ASHRAE recommended value for CO_2.

Proportion of PM fractions

To understand the contribution of different PM sizes from various indoor and outdoor sources, fractional analysis of Particulate matter has also been done. The proportion of various PM fractions in each of the chosen indoor microenvironments is displayed in Figure 3.
Approximately 59% of the total PM_{10} in homes comes from PM_1 (45%) and $PM_{2.5}$–PM_1 (14%). This indicates a dominance of fine particles likely generated from indoor sources such as cooking, use of household products, and limited mechanical ventilation. In office spaces, the fractional composition shows PM_1 ~38% and $PM_{2.5}$–PM_1 ~17%, because of better-controlled conditions with mechanical ventilation systems and lower levels of indoor activity, the combined percentage of PM_1 and $PM_{2.5}$–PM_1 in offices is slightly lower at 55%. In contrast,

classrooms exhibit a distinct pattern, with just 44% of PM_{10} being made up of PM_1 (28%) and $PM_{2.5}$–PM_1 (16%). This higher share of coarse particles (56%) could be due to higher occupancy density, frequent movement, and dust resuspension from floors and surfaces in the absence of adequate air filtration systems[6]. These differences highlight the influence of indoor functional use, occupancy patterns, and ventilation conditions on particulate matter composition.

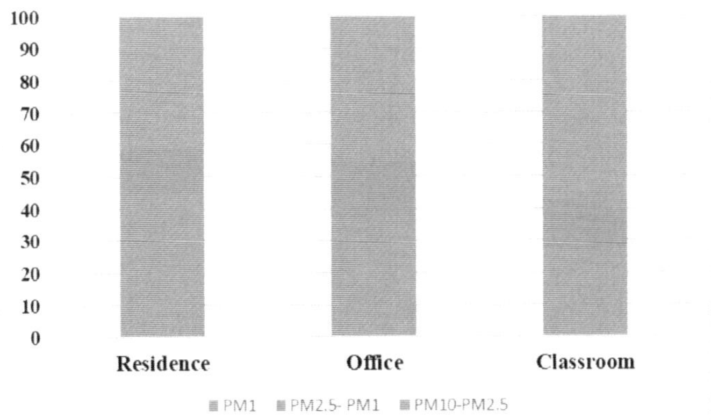

Figure 3. Proportion of PM concentrations in various size ranges.

4 DISCUSSION

Distribution of indoor pollutants (PM_{10}, $PM_{2.5}$, PM_1 and CO_2 levels) was examined for each of the selected microenvironment. Statistical analysis confirms the spatial distribution of pollutants across the indoor environments. The peaks observed in PM concentrations coincide with the routine activities like cooking and cleaning, confirming these as the major sources of indoor particles. The CO_2 concentration patterns further support the impact of occupancy and activities, peaking during periods of high occupancy, especially in classrooms and offices.

The classroom environment demonstrated the highest PM_{10} levels (~472.6 μg/m³), likely due to high movement-induced resuspension of coarse particles, a finding in agreement with previous literature on school indoor air pollution.[7]. Conversely, offices showed comparatively lower PM concentrations but higher CO_2 levels (~816.8 ppm), indicative of prolonged occupancy combined with inadequate ventilation, a common issue in modern office buildings. Fractional analysis revealed that fine particles (PM_1 and $PM_{2.5}$–PM_1) dominate residential settings, consistent with indoor combustion and household product use, while classrooms had a greater proportion of coarse particles, suggesting dust resuspension as a major contributor. However, several potential limitations may influence the interpretation of these results. Measured pollution levels may be impacted by short-term monitoring intervals, inaccurate sensor calibration, or unreported external infiltration. Furthermore, it is more difficult to identify precise pollution sources when there is less control over occupant behaviour and ventilation settings during sampling. Despite the limitations, the robustness of the results is supported by the correspondence between recorded occupant activities and observed temporal patterns.

5 CONCLUSIONS

This study underscores the critical role of occupant behaviour, ventilation efficiency, and indoor activities in shaping indoor air quality across different microenvironments. The significantly higher particulate matter levels in homes and classrooms indicate that better air quality management techniques are required in these environments, especially through better ventilation and pollution source control. On the other hand, peak concentrations in office space indicate that even controlled facilities can experience insufficient ventilation during periods of high occupancy, even if CO_2 levels generally stayed below suggested standards.

These insights have practical implications for researchers, building designers, facility managers, and occupants. For example, incorporating real-time IAQ monitoring and adaptive ventilation controls can mitigate pollution peaks associated with specific activities. In order to provide comprehensive, site-specific IAQ solutions that protect occupant productivity and health and to develop standard IAQ guidelines for various indoor settings in India, further study may be performed on additional indoor environments and climatic situations.

6 REFERENCES

[1] S. A. Ahmed Abdul–Wahab, S. C. F. En, A. Elkamel, L. Ahmadi, and K. Yetilmezsoy, "A review of standards and guidelines set by international bodies for the parameters of indoor air quality," *Atmos Pollut Res*, vol. 6, no. 5, pp. 751–767, Sep. 2015, doi: 10.5094/APR.2015.084.

[2] T. Schneider, "Dust and fibers as a cause of indoor environment problems," 2008.

[3] R. Goyal and M. Khare, "Indoor-outdoor concentrations of RSPM in classroom of a naturally ventilated school building near an urban traffic roadway," *Atmos Environ*, vol. 43, no. 38, pp. 6026–6038, 2009, doi: 10.1016/j.atmosenv.2009.08.031.

[4] GRIMM Aerosol Technik GmbH & Co, KG, User Manual for Portable Laser Aerosol spectrometer and Dust Monitor, GRIMM Aerosol Technik GmbH & Co. KG, Ainring, 2010. https://www.wmo-gaw-wcc-aerosol-physics.org/files/opc-grimm-model– 1.108-and-1.109.pdf

[5] P. Kapalo *et al.*, "Investigation of CO2 production depending on physical activity of students," *Int J Environ Health Res*, vol. 29, no. 1, pp. 31–44, Jan. 2019, doi: 10.1080/09603123.2018.1506570.

[6] S. M. Almeida *et al.*, "Children exposure to atmospheric particles in indoor of Lisbon primary schools," *Atmos Environ*, vol. 45, no. 40, pp. 7594–7599, Dec. 2011, doi: 10.1016/j.atmosenv.2010.11.052..

[7] M. Elbayoumi, N. A. Ramli, N. F. F. Md Yusof, and W. Al Madhoun, "Spatial and seasonal variation of particulate matter (PM10 and PM2.5) in Middle Eastern classrooms," *Atmos Environ*, vol. 80, pp. 389–397, Dec. 2013, doi: 10.1016/j.atmosenv.2013.07.067.

Investigating the Relationship between Energy Use and IEQ in Tropical Office Building: Enhancing the quality of EEE for Retrofit Solutions.

Abha Durgesh * , **K Thirumaran**
Department of Architecture, National Institute of Technology, Tiruchirappalli

Optimum IEQ enhances health, comfort, productivity, and sustainability but often depends on energy-intensive mechanical systems. With rising energy demands and ongoing energy crises, it is essential to reduce energy consumption without compromising IEQ. This research aims to bridge the gap between energy consumption & IEQ, particularly in tropical settings where cooling demands are high.

This study investigates the Environmental Energy Efficiency (EEE) value as an effective tool in the building retrofitting process to assess the combined improvement in IEQ and Energy efficiency.

The study was conducted in an office building located in Tiruchirappalli, Tamil Nadu. Various spaces within the building were selected based on their frequency of use and subjected to a detailed analysis. The EEE values for these selected spaces were calculated using results from an EnergyPlus model, which was validated through on-site field measurements. Subsequently, several retrofitting solutions were proposed in alignment with the ECBC after comparing the existing energy use, thermal comfort, and visual comfort against their respective benchmarks. The retrofitted building was re-simulated, and the EEE value was recalculated and compared with the baseline to evaluate improvements.

The results show a significant improvement from the Base Case EEE to the Retrofit Case EEE, indicating notable advancements in environmental quality and energy efficiency.

The study illustrates how EEE values quantify the benefits of retrofit strategies, aiding in justifying investment in proposed retrofitting. Further research could integrate cost considerations into this process, facilitating more informed decision-making for clients considering building retrofits.

Keyword: *Energy Efficiency, EEE, Building Retrofit*

Investigating the Relationship between Energy Use and IEQ in Tropical Office Building: Enhancing the quality of EEE for Retrofit Solutions.

Abha Durgesh[1],*,Dr. K.Thirumaran[1],

1. Department of Architecture, National Institute of Technology, Tiruchirapalli, India.

Corresponding email: durgeshabha@gmail.com

SUMMARY

Enhancing IEQ improves health, comfort, productivity, and sustainability, but often depends on energy-intensive mechanical systems. This study investigates the Environmental Energy Efficiency (EEE) value as an effective tool in the building retrofitting process to assess the combined improvement in IEQ and Energy efficiency.

Various spaces within the office building in Tiruchirappalli, Tamil Nadu, were analysed using Energy Plus simulations, validated with field data. Subsequently, several retrofitting solutions were proposed in alignment with the ECBC after comparing the existing energy use, thermal comfort, visual comfort, and IAQ against their respective benchmarks. The retrofitted building was re-simulated, and the EEE value was recalculated and compared with the baseline to evaluate improvements.

The findings show a significant improvement from the Base Case EEE to the Retrofit Case EEE, indicating notable advancements in environmental quality and energy efficiency. Future work could incorporate economic analysis to aid clients in making more informed choices regarding sustainable building upgrades.

KEYWORDS

EEE, IEQ, Energy Efficiency, Building Retrofit.

1. INTRODUCTION

The buildings and construction sector accounts for a substantial portion of global energy use and greenhouse gas emissions, with material production and operational activities contributing approximately 37% of energy demand and 34% of CO_2 emissions (IEA, 2023). As the climate crisis intensifies, enhancing energy performance in both new and existing buildings has become a central strategy for global decarbonization (UNEP, 2024).

With people spending nearly 90% of their time indoors, ensuring optimal indoor environmental conditions, such as thermal, visual, air quality, and acoustics, is vital for enhancing workplace health, comfort, and performance (Fissore, 2023).

Energy-efficient buildings not only reduce energy consumption but also improve IEQ, enhancing occupant comfort and health (Geng Y. L., 2019).In tropical climates, balancing energy efficiency and IEQ is challenging due to high thermal loads (Verma, 2023). Thus, balancing energy use and IEQ is critical to ensure improvements in one do not compromise the other (Roumi, 2023) .

Retrofitting as a key strategy to improve energy efficiency and reduce environmental impact in tropical climates (Chung-Camargo, 2024). Deep energy retrofits-such as upgrading insulation, HVAC systems, lighting, and the building envelope can reduce energy use by 30-50% (Sharma, 2014) (Ascione, 2016). However, global retrofit rates remain low, around 1% per year, which is insufficient to meet climate objectives for 2030 and 2050 (IEA, 2023). Frameworks like the Environmental Energy Efficiency (EEE) index have been proposed to help overcome these challenges by integrating performance data to guide retrofitting strategies (Yang Geng, 2020).

2. METHODOLOGY

The study followed a structured process including planning and preparation, data collection, data analysis and calculation of the Environmental Energy Efficiency (EEE) index. This approach was applied to various spaces based on frequency of users in the Training and Placement (CAPSTONE) Building at the NITT, which is situated in a tropical climate.

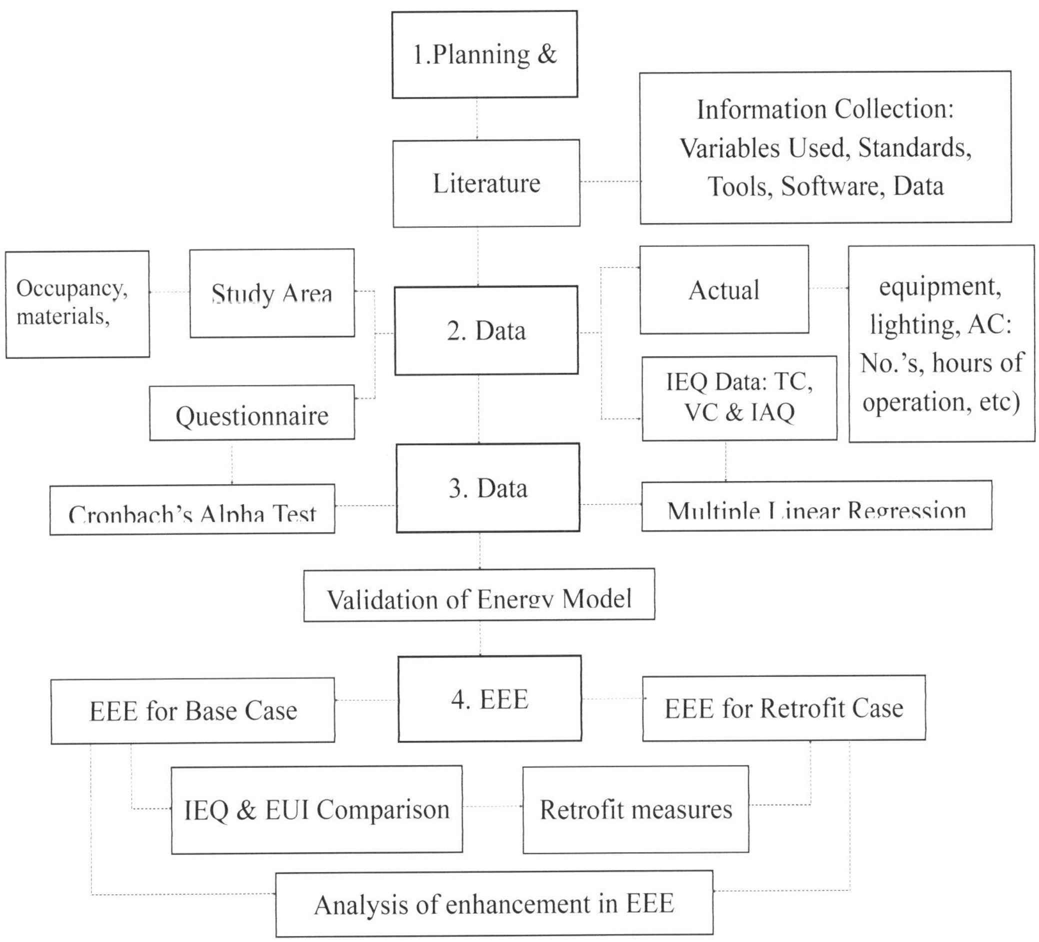

Figure 1. Methodology framework

The IEQ assessment considered thermal comfort air temperature (AT) and relative humidity (RH) , visual comfort (illuminance), and indoor air quality (CO_2 concentration), following the methodology outlined by (Sansaniwal, 2023). Measurements were taken using the Heat Stress WBGT meter for AT (0-50 °C) and RH ((1%–99%) with ±0.1 °C and ±3–5% accuracy, an air quality meter for CO_2 (400–5000 ppm, ±5% accuracy), and the HD450 Light Meter for illuminance (0–400,000 lux, ±5% accuracy). Questionnaire reliability is assessed using Cronbach's alpha test using IBM-SPSS statistics, and MLR analysis variable relationship.

Table 1 Study Area Details

S.no.	Space Selection	Area	Location	Image	Measurement approach
1	Interview Room 1	15.4 sq. mt.	Ground floor; East Facing		
2	Interview Room 2	15.4 sq. mt.	First floor; East Facing		
3	Interview Room 3	15.4 sq. mt.	First Floor; Courtyard side (west)		
4	Office	108 sq. mt.	Ground Floor; West facing		
5	Pre-Presentation Talk Room (PPT)	113.6 sq. mt.	Ground floor; South facing		

2.1 EEE CALCULATION

The Environmental Energy Efficiency (EEE) index was introduced by (Yang Geng, 2020) to assess building performance by integrating both normalised IEQ performance and energy consumption. This index combines objective data and subjective occupant feedback, providing a comprehensive yet data- demanding evaluation. Energy consumption is normalised against standard benchmark based on building type and climate zone. The detailed calculation procedure is provided in Eqs. (1) – (3).

$$\text{Environmental Energy Efficiency(EEE)} = \frac{\text{Normalized IEQ Performance}}{\text{Normalized Building Energy}}$$

(1)

$$\text{Normalized IEQ Performance} = \sum_{i=1}^{n} \alpha_i . [\, \beta_i . \text{IEQ}_{i,ob} + (1 - \beta_i) . \text{IEQ}_{i,sub}]$$

(2)

$$\text{Normalized Building Energy} = \frac{\text{Actual Energy Use}}{\text{Energy constraint value of corresponding building}}$$

(3)

In eq (2), n=3 IEQ factors (thermal comfort, air quality, visual comfort). $\text{IEQ}_{i,ob}$ denotes compliance rate of measured parameters, while $\text{IEQ}_{i,sub}$ is the user satisfaction ratio. α_i indicates the weight of each factor and β_i reflects the weight of objective performance (Yang Geng, 2020). Based on (Francesco Leccese, 2021), α values are 0.52 (TC), 0.21 (VC), and 0.27 (IAQ), with β = 0.33 for all. As per (BEE) standards, the EPI benchmark for office building in warm and humid climate is 182 kWh/m^2/year.

3. RESULT

The energy model demonstrates strong accuracy (RMSE 0.76, CVRMSE 2.76%, MBE 0.56), confirming reliability for energy predictions.

In both base case and retrofit cases, people density is 0.19p/sq. m (interview room), 0.27p/sq. m (office), and 0.44 p/sq. m (pre-presentation talk). Equipment power density is 4 W / sq. m (Interview room), 15 W/sq. m (office), and 4.3 W/ sq. m. (PPT). the temperature set points range from 24.04°C to 27.04 °C. Lighting power density is 7 W/ sq. m. (Interview room), 1.33 W/sq. m (Office), and 8.23 w/sq. m. (PPT) in the base case, and 3.5 W/sq. m (Interview room), 6 W/ sq. m (office) and 6 W/ sq. m (PPT) in the Retrofit Case.

The retrofit case shows significant improvement to meet ECBC guidelines by achieving daylight requirements, reducing VLT to 0.78 with low-e-glazing, and lowering U-values: 1.5 W/m^{2}K (fenestration), 0.37 W/m^{2}K (walls), and 0.39 W/m^{2}K (roof). WWR remained at 47%, external shading was unchanged, and overcast sky conditions were considered.

3.1 BASE CASE EEE CALCULATION

IEQ components were evaluated using an MLR model with 55 survey results. Interview room at ground floor (0.31,0.59,0.38), interview room at first floor (0.29,0.50,0.37), Office space (0.20,0.40,0.35), PPT space (0.35,0.53,0.37) were the subjective ratios for TC, VC, and IAQ.

Base case compliance ratios demonstrated that interview rooms performed better regarding IEQ than PPT and Office environments. TC= 0.71, IAQ= 028, and VC=0.75 were recorded in the ground floor interview room. In contrast, the first-floor (east) and the courtyard side interview rooms displayed comparable TC (0.73 & 0.71) and IAQ (0.30), with VC (0.75 and 0.66) respectively. The office's compliance was lower (TC=0.66, IAQ=0.28, VC=0); the PPT space's compliance was moderate (TC=0.67, IAQ=0.40), but no VC compliance.

The base case energy usage for different areas, the office uses 30,371 kWh, for an EUI of 281.21 kWh/m²/yr, which is more than the BEE guideline of 182 kWh/m²/yr. Using 24,471 kWh, the PPT space has an EUI of 215.41 kWh/m²/yr. Using 2,977 kWh, the ground floor interview room has an EUI of 193.31 kWh/m²/yr. The first-floor interview rooms exhibit comparable patterns: the courtyard side interview room uses 3,560 kWh (EUI=231.16 kWh/m²/yr) and the east-facing interview room consumes 3,789 kWh (EUI=246.03 kWh/m²/yr).

The base case Energy Efficiency Evaluation (EEE) values indicate higher efficiency in the interview room at ground floor is 0.57, followed by PPT space 0.49, interview room at courtyard side (west) is 0.48, interview room at first floor east facing is 0.46, and the office space showing the lowest value of 0.35.

3.2 RETROFIT CASE EEE CALCULATION

The retrofit case energy usage for different areas, the office uses 25,357 kWh, for an EUI of 234.7 kWh/m²/yr. Using 18771 kWh, the PPT space has an EUI of 165.2 kWh/m²/yr. Using 1708 kWh, the interview room 1 has an EUI of 110.9 kWh/m²/yr. Using 2515 kWh, the interview room 2 has an EUI of 163.3 kWh/m²/yr, and using 2797 kWh, the interview room 3 has an EUI of 181.6 kWh/m²/yr.

In the retrofit case, Interview Room 1 shows compliance ratios of 0.80 for thermal comfort (TC), 0.26 for indoor air quality (IAQ), and 1 for visual comfort (VC). Interview room 2 records 0.86(TC), 0.28 (IAQ, and 1 (VC), while interview room 3 has 0.82 (TC), 0.28 (IAQ), and 1 (VC). The office space achieves 0.72 (TC), 0.06 (IAQ), and 0.9 (VC), and the PPT space records 0.71 (TC), 0.39 (IAQ), and 1 (VC). In the retrofit case, taking into account 90 % satisfaction (IMAC)for the subjective ratio of TC, VC, and IAQ while fulfilling the requirements given in (Sansaniwal, 2023).

The retrofit case EEE in interview room 1 is highest at 1.38, followed by interview room 2 at 0.94, PPT space at 0.83, interview room 3 at 0.82, and the lowest is in the office space at 0.62.

4. DISCUSSION

Significant improvements are evident when comparing the Retrofit Case EEE for various spaces with the Base Case EEE. A 1.42 times enhancement in EEE in interview room 1 is clear. In interview room 2 EEE, the enhancement is 1.04 times, and in interview room 3 EEE, the enhancement is 0.7 times. In office space, the EEE enhancement is 0.77 times. In the PPT space, a 0.69 times EEE enhancement is evident.

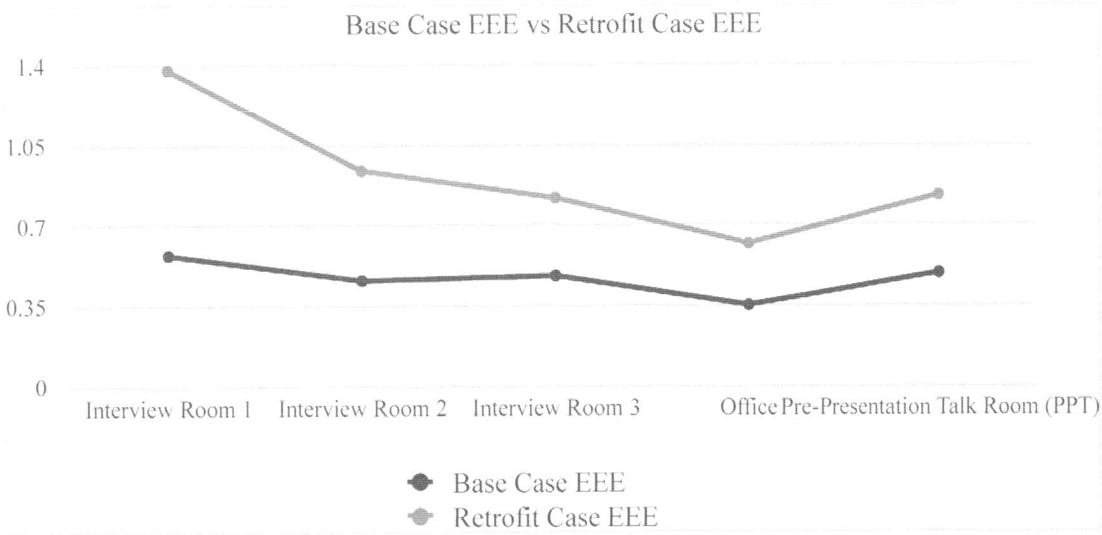

Figure 2 EEE comparison of Base Case with Retrofit Case

5. CONCLUSION

The environmental performance analysis of the Capstone Building reveals notable gains in energy efficiency and Indoor Environmental Quality (IEQ) as a result of retrofit interventions. Increased occupant well-being and productivity were a result of improvements in thermal comfort, visual comfort, and indoor air quality, which were made possible by better building envelopes, upgraded lighting, and optimised HVAC systems. Higher comfort hours, improved illumination, lower Energy Use Intensity (EUI), and compliance with IMAC and ISHRAE standards were all results of retrofit procedures. Notably, significant improvements in environmental quality and energy efficiency were shown while switching from the Base Case EEE to the Retrofit Case EEE. These results support the efficacy of retrofit strategies. Future research can further assist clients seeking building retrofits in making well-informed decisions by incorporating cost analysis.

6. ACKNOWLEDGEMENT

We appreciate the Department's provision of research facilities and the participation of NIT Tiruchirapalli students and staff in the survey.

7. REFERENCES

al, W. e. (2023). Assessment and simulation of evacuation in large railway stations. *Building and Environment*.

Ascione, F. d. (2016). Energy and environmental retrofitting of existing buildings: Energy efficiency potential of deep retrofits in Europe. . *Energy and Buildings*.

BEE. (n.d.). *Bureau of Energy Efficiency.*

Chung-Camargo, K. e. (2024). Advances in Retrofitting Strategies for Energy Efficiency in Tropical Climates: A Systematic Review and Analysis. *Buildings*.

Fissore, V. I. (2023). Indoor Environmental Quality and Comfort in Offices: A Review. *MDPI*.

Francesco Leccese, M. R. (2021). Towards a holistic approach to indoor environmental quality assessment: Weighting schemes to combine effects of multiple environmental factors. *Energy and Buildings*.

Geng, Y. L. (2019). Energy-efficient buildings and their impact on indoor environmental quality (IEQ): Enhancing occupant comfort and health. *Journal of Environmental Management*.

Geng, Y. L. (2020). A review of operating performance in green buildings: Performance data to inform effective retrofitting strategies. *Journal of Cleaner Production*.

IEA. (2023). *World Energy Outlook.*

IMAC. (n.d.). *IMAC Assistant (India Model For Adaptive Thermal Comfort Tool Assistant).* CARBSE.

Ralegaonkar, V. S. (2014). Indoor environmental quality: review of parameters and assessment models. *Taylor & Francis*.

Roumi, I. T. (2023). Balancing energy use and indoor environmental quality in office buildings: A study on energy-efficient retrofitting strategies. *Journal of Building Performance*.

Sansaniwal, S. K. (2023). Development and validation of ISHRAE Indoor Environmental Quality Standard. *Research Gate*.

Sharma, A. C. (2014). Deep energy retrofitting strategies for existing buildings: A case study in India. *Renewable and Sustainable Energy Reviews*.

UNEP. (2024). *United Nations Environment Programme (UNEP) and Global Alliance for Buildings and Construction (GlobalABC).*

Verma, A. e. (2023). Tropical climates and the interplay between IEQ and energy consumption in buildings: A review. *Building and Environment*.

Yang Geng a, b. ,. (2019). A review of operating performance in green buildings: Energy use, indoor environmental quality and occupant satisfaction. *Energy & Building*.

Yang Geng, B. L. (2020). Comparative study on indoor environmental quality of green office buildings with different levels of energy use intensity . *Building and Environment*.

Enhancing Indoor Air Quality in Healthy Buildings: The Role of Advanced Ventilation Systems

Deepak Nanaware *

American Air Filter, India, Bangalore

As urbanization and indoor air quality concerns continue to rise, ensuring healthy indoor environments is crucial for occupant well-being and productivity. This paper explores the significant role of advanced ventilation systems in enhancing indoor air quality (IAQ) in modern buildings. It examines how building architecture design and mechanical, electrical, and plumbing (MEP) systems influence IAQ, highlighting both their positive impacts and inherent limitations. The paper emphasizes the need for adequate fresh air supply, discussing the challenges of balancing energy efficiency with sufficient ventilation. Additionally, it explores the sources of internal and external pollutants, such as building materials, occupant activities, and outdoor air contaminants, and their implications for health. By analysing the interplay between design, systems, and pollutants, this study aims to provide insights into optimizing ventilation strategies that foster healthier, more sustainable indoor environments.

Keyword: *Indoor Air Quality, Indoor Air Pollution, Resident Health*

Emissions of volatile organic compounds from cleaning, disinfecting and care products used in daycare centers from the French CRESPI cohort

Mélanie Nicolas[1], Pierre Bonnet[1], Orianne Dumas[2], Doriane Rousselle[1], Priscilla Thiry[1], Nicole Le Moual[2], Corinne Mandin * [3]

[1] *Scientific and Technical Center for Building, France*

[2] *Université Paris-Saclay, UVSQ, Univ. Paris-Sud, Inserm, Équipe d'Épidémiologie Respiratoire Intégrative, CESP Villejuif, France*

[3] *Nuclear Safety and Radiation Protection Authority, France*

Indoor air quality (IAQ) in daycare centers, where young children spend a large amount of time, is of utmost importance. Improving IAQ relies on the knowledge of emission sources to reduce them. In daycare centers, cleaning and disinfecting products are used daily but their indoor emissions have rarely been evaluated. In this context, a collection of the cleaning, disinfecting and care products was carried out in daycare centers in the frame of the CRESPI cohort.

Emissions of volatile and semi-volatile organic compounds (VOC and SVOC) used in 102 daycare centers were assessed in experimental test chambers in accordance with ISO 16000 and ISO 16516+A1 standards. Realistic application scenarios were set up and the air chamber was sampled during three consecutive 30-minute periods after each product use, for the analysis of aldehydes, VOCs and SVOCs. For each emitted compound, the hazard classifications were retrieved from various databases.

Out of a total of 223 unique commercial references of products used in daycare centers, 70 products (30 cleaning, 16 disinfecting, 19 care and 5 hydroalcoholic products) were assessed. 442 compounds were detected at a concentration above 1 µg/m^3 toluene equivalent. Aldehydes were the most frequently emitted compounds, while terpenes and alcohols were emitted in the highest concentrations. Regarding hazard classifications, 5 compounds are classified as carcinogenic, 6 as mutagenic, 8 as toxic to reproduction, 37 as endocrine disruptors, 7 with chronic toxicity, and 5 as skin sensitizers. These results confirm the need to limit the use of products to what is strictly necessary.

Keyword: *IAQ, VOCs, cosmetics*

Thermal Comfort Assessment and Optimization in Residential Buildings: Integrating Vastu Shastra Principles with Advanced Simulation Tools

ANURAG AMAN KAUSHAL * [1], **Prashant Anand**[2], **Bharath Haridas Aithal**[3]

[1] *Centre of Excellence for Indian Knowledge Systems, Indian Institute of Technology Kharagpur*

[2] *Department of Architecture and Regional Planning, Indian Institute of Technology Kharagpur*

[3] *RCG School of Infrastructure Design and Management, Indian Institute of Technology Kharagpur*

This study integrates traditional architectural principles, specifically Vastu Shastra and Dandaka planning, with simulation tools to understand indoor thermal comfort in residential buildings. The research evaluates the impact of door and window orientations on thermal comfort during summer, focusing on living rooms representing a cardinal orientation with varying door and window opening sizes in different buildings. Thermal comfort indices—Predicted Mean Vote (PMV), Predicted Percentage Dissatisfied (PPD), and Standard Effective Temperature (SET)—were calculated via the CBE thermal comfort tool based on the ASHRAE thermal comfort model. Field measurements validated computational models, including temperature, humidity, and airflow. The results highlighted the benefits of the layouts and orientation based on Vastu shastra, which exhibit lower PMV values due to natural shading and enhanced airflow, maintaining stable indoor temperatures and minimizing dependence on mechanical cooling. The study identified significant variations in comfort across orientations. West-facing living rooms exhibited the highest average PMV of 2.0, indicating warmth and discomfort, while east-facing rooms had an average PMV of 1.3, suggesting slight warmth within comfort limits. Building simulations using Rhino, Ladybug, and Honeybee validated these findings, highlighting how orientation and spatial configurations influence thermal conditions. Quantitatively, the results demonstrated a 20% reduction in cooling loads due to these passive strategies. These findings validate the effectiveness of Vastu Shastra's principles in optimizing indoor thermal conditions. They underscore the potential of combining traditional design wisdom with modern computational tools to create a thermally comfortable living environment.

Keyword: *Vastu Shastra, Thermal Comfort, Building Simulations*

Indoor Air Quality in a Delhi School: Data-Driven Insights from a Monitoring Study

Lakshmy N *, **Abdullah Kidwai, Rajeev Kumar, Nikunj Shukla, Ashish Gupta, Nidhi Gupta, Piyush Varma, Gurneet Singh**

Environmental Design Solutions, New Delhi, India

Children are particularly susceptible to the adverse health effects of air pollution due to their developing organs, higher breathing rates, and increased air inhalation. Given that they spend significant time in schools, assessing indoor air quality (IAQ) is crucial. This study monitored carbon dioxide (CO_2), particulate matter (PM_{10}, $PM_{2.5}$), total volatile organic compounds (TVOCs), relative humidity (RH), and temperature in classroom, library, computer lab, and chemistry lab in a school in Delhi.The study was designed to answer the following research questions: (a) duration for which the IAQ parameters remained within recommended thresholds; (b) correlations between indoor and outdoor PM concentrations; and (c) impact of daily activities, building design, and operational policies on these parameters. The study revealed that $PM_{2.5}$ and PM_{10} levels met WHO interim thresholds for 55% and 47% of the time, respectively. TVOC and CO_2 levels adhered to the WELL standard for 49% and 95% of the time, respectively. Indoor PM levels directly correlated with respective outdoor levels. The school being designed and operated for natural ventilation contributed to this direct correlation between indoor and outdoor PM levels, dissipation of TVOCs during the day and safe levels of CO_2. Poor ambient PM levels and activities like dry dusting, use of chalks, chemical experiments contributed to increased PM levels indoors. Nighttime off-gassing of TVOCs was attributed to materials like books, stationery, computers and cleaning chemicals. The study highlights the need for targeted indoor and outdoor mitigation strategies to improve indoor air quality in schools.

References:

www.unicef.org. (n.d.). Air pollution: A threat to children's rights | UNICEF. [online] Available at: https://www.unicef.org/stories/air-pollution-threat-childrens-rights.

Keyword: *IAQ Monitoring, School, Carbon Dioxide, Particulate Matter, Volatile Organic Compounds*

Critical review of literature assessing indoor air quality (IAQ) in green buildings

Udita Gupta * , **Wendy Miller, Richard Brown, Lidia Morawska**
The ARC Training Centre for Advanced Building Systems against Airborne Infection Transmission (THRIVE), Queensland University of Technology, Brisbane, QLD, 4000, Australia

Less than a tenth of the weighting towards green building certification (GBC) comes from indoor environmental quality parameters (IEQ) of which, indoor air quality (IAQ) is a subcategory. The overall weight given to "IAQ" in buildings is therefore effectively negligible. A systematic PRISMA literature review was conducted to answer the question "Do green buildings have good/satisfactory IAQ?" A search of four databases, Web of Science, Science Direct, Scopus, and Taylor and Francis produced 47, 194, 92 and 248 papers (total 581) respectively, which were narrowed down to 85. These were classified into three categories: A: review/opinion paper (38); B: comparison of IAQ/IEQ in green and conventional buildings (24), and C: studies based on occupant surveys (26). Some of those papers were classified in both categories B and C. Of these (B, C), 14 studies concluded green buildings have good air quality, 9 studies denied green buildings have good IAQ, 12 studies found contradicting results, and 5 studies clearly state that "no significant difference is observed in these two types of buildings". This review is spread across all GBC with indoor air quality requirements and includes LEED, KGBC, Minergie, Green Star, BREEAM, Home Stars, CASBEE, IGBC, Pearl, GPRS, Green Globes, Beam Plus, SI 5281, KLIMA, HQE. IAQ/IEQ studies were found only for first five certifications and were dominated by LEED certification. The findings emphasize the urgent need for conducting post-commissioning IAQ performance monitoring, ensuring buildings are equipped with reasonable air-quality-control facilities and improve guidelines based on its feedback. ***This abstract was supported by the ARC IC220100012 THRIVE***.

Keywords: "Sustainable buildings", "post occupancy evaluation", "indoor air pollution", "carbon dioxide", "ventilation"

Comprehensive analysis of an Informal Settlement for Thermal Comfort.

Dhanashree Yadav *, **Namrata Dhamankar, Dr. Prajakta Kulkarni**
MKSSS Dr. Bhanuben Nanavati College of Architecture for Women, Savitribai Phule Pune University.

Over 50% of the global population now resides in urban areas, with 90% of future urban growth expected in Asian and African cities, placing immense pressure on infrastructure, services, and the environment. A key consequence of this expansion, particularly in emerging economies, is the rise of informal settlements, which house a significant portion of the urban poor and are highly vulnerable to climate impacts. According to a 2023 UN-Habitat report, nearly 30% of urban residents live in informal settlements. While extensive research has explored socio-cultural aspects, encroachment issues, and sanitation, studies on the built environment and environmental comfort in these areas remain limited. In Indian cities, high demand for indoor cooling arises due to climatic conditions. Informal settlement dwellings, built with unconventional materials, often create uncomfortable indoor environments. This study builds upon the research titled "Marginalised Futures of Sustainable Cooling in Pune," conducted by the University of Leeds, UK, and BNCA, which used a co-production methodology to explore the lived experiences of heat stress and cooling needs in Pune's informal settlements. While that research primarily focused on qualitative data, the quantitative aspect remained underexplored. To address this gap, the present study undertakes a detailed quantitative analysis of thermal comfort, contributing a data-driven perspective to the existing knowledge. The study analyses onsite readings and observations of homes in informal settlements to deepen understanding of climate-related challenges. Focusing on Pune, it examines indoor environmental quality using data loggers and household surveys to assess behavioural patterns. Air temperature, mean radiant temperature, humidity, air speed, and air quality index (AQI) were monitored. Fifty residents participated in oral surveys. Twelve households were assessed, with three selected for detailed environmental data collection. The findings aim to support strategies that improve thermal comfort in informal urban communities.

References:

1. Sarbu, I. (2013). Aspects of indoor environmental quality assessment in buildings. Journal of Energy and Buildings.
https://www.sciencedirect.com/science/article/abs/pii/S037877881300073X?via%3Dihub
2. https://unhabitat.org/sites/default/files/2024/05/annual_report_2023.pdf
3. Elsayed M. (2023) Post-occupancy evaluation in residential buildings: A systematic literature review of current practices in the EU. Journal of Building and Environment. Post-occupancy evaluation in residential buildings: A systematic literature review of current practices in the EU - ScienceDirect.

Keyword: *Urban expansion, thermal comfort, Urban Heat*

Design Considerations for Developing Compact and Low-Cost Light Scattering-based PM IoT Devices.

Roshan Yadav * [1], **Rajesh Zele**[1], **Abhishek Chakraborty**[2]
[1] *Department of Electrical Engineering, Indian Institute of Technology Bombay, India*
[2] *Environmental Science & Engineering Department, Indian Institute of Technology Bombay, India*

Indoor Air Quality (IAQ) is a critical health threat, with approximately 2.1 billion people still using solid fuels and kerosene for cooking, resulting in inhaling air with high levels of particulate matter, especially in poorly ventilated homes. Real-time monitoring of particulate matter can play a significant role in understanding and reducing these health risks. This paper presents design considerations for developing a low-cost, low-power multi-PM sensor-based IoT device for real-time IAQ monitoring. In the proposed approach, a real-time operating system algorithm simultaneously collects data from commercial sensors, resulting in time-synchronized data acquisition stored in the local memory. It has embedded Wi-Fi connectivity, which enables online data sharing for visualization. The entire system is enclosed in a 3D-printed case designed with the self-heating and airflow of the sensors in mind. The results have been validated in an indoor setting against a high-quality reference-grade analyzer with each sensor, yielding coefficient of determination (r^2) values > 0.70. The device is compact, measuring $14 \times 7 \times 8$ cm^3, and has a runtime of over 4 days on a single charge, costing less than USD 100, offering a practical and efficient option for real-time IAQ monitoring.

Keyword: *Indoor Air Quality, Particulate Matter, Internet of Things*

A Multidisciplinary Institutional Approach to Study IAQ in Classrooms to Mitigate Airborne Disease Transmission Risk

Pushpesh Singh[1], Ravinder Arigela[2], Srinidhi Balasubramanian[2,3], Krishnendu Sinha [*1]

[1] *Department of Aerospace Engineering, Indian Institute of Technology Bombay, Mumbai, India*

[2] *Environmental Science and Engineering Department, Indian Institute of Technology Bombay, Mumbai, India*

[3] *Center for Climate Studies, Indian Institute of Technology Bombay, Mumbai, India*

Indoor environmental quality (IEQ) plays a critical role in determining the health, well-being, and cognitive performance of occupants, especially in high-occupancy environments such as classrooms. Here, we showcase institutional-level efforts recently undertaken at the Indian Institute of Technology Bombay to characterize several aspects of the indoor environment and approaches for improving the IEQ in classrooms.

A series of independent yet interconnected research projects were developed to characterize airflow, CO_2, particulate matter, and bioaerosols, identify optimal placement of sensors for representative measurement, and develop models to characterize airborne infection risk. These studies were conducted at multiple scales including a reduced scale model of a classroom for bio testing, live classrooms of different dimensions and occupancy, and controlled experiments in a newly developed, dedicated life-size ventilation testbed. Based on the collective findings, a long-term IoT-driven monitoring and intervention plan is under deployment to characterize IEQ in three representative classrooms with varying occupancy and geometry. Practical implementation challenges, such as sensor robustness, optimal sensor placement, and HVAC constraints, are being addressed.

Representative results covering the different aspects of modeling, simulation, experimental and live data from the classroom will be presented. These will be used to highlight the challenges in classroom ventilation and propose interventions to mitigate air-borne infection risk, improve IEQ and develop energy-efficient solutions for a cleaner healthier learning environment.

Keyword: *Classroom, Ventilation, Airborne infection*

Passive and Active Solutions for Formaldehyde Mitigation in Office Buildings

E V S Kiran Kumar Donthu * [1], **Shivraj Dhaka**[2], **Tanya Kaur Bedi**[3]

[1] *The Energy & Resources Institute, India & Indian Institute of Technology (IIT)-BHU, Varanasi, India*

[2] *Indian Green Building Council (IGBC), India*

[3] *School of Planning and Architecture, Bhopal*

Formaldehyde (HCHO) is one of the significant indoor air pollutants that reacts and behaves dynamically in response to indoor environmental conditions. Exposure to HCHO can lead to short term and long-term human health issues. Therefore, lowering HCHO is an important objective of constructing healthy buildings. In this study, the experimental observation of two passive and active indoor HCHO reduction strategies have been demonstrated in two office buildings located in New Delhi and Bangalore. HCHO levels were monitored at 30-minute interval on consecutive days using portable photoelectric absorptiometric sensors along with the indoor environmental conditions such as temperature and relative humidity. These measured levels were compared with and without 1) passive treatments that encourages dilution through, indoor plants and a wall paint, and 2) active solutions, such as a mechanical ioniser. For the indoor plants, pollution abatement paint, and mechanical ioniser, the reduction in the 8-hour average of HCHO levels was determined to be 11.25 ppb (23%), 10.5 ppb (42.2%), and 7.4 ppb (27.4%), respectively. Passive solutions such as selective indoor plants and pollution abatement paints seem to be as effective as active solutions in mitigating HCHO as was observed in this study. Passive solutions gain more significance against active solutions especially in the context of designing healthy buildings. This research sheds some light on the implementation of these strategies in office buildings. The limitation, though, is that it doesn't investigate whether using these mitigation strategies could result in an increase in secondary pollutants.

Keyword: *Formaldehyde, Mitigation, Paints*

“Evaluating the Role of School Environmental Quality factors in Shaping Student Health & wellbeing and Parental Preferences: A Comparative Study of Private and Government Schools”

Mansi Mokashi *, **Sujata Karve**
Dr. Bhanuben Nanavati College of Architecture, Pune, India

ABSTRACT

This research investigates the impact of School Environmental Quality on student health and well-being and explores whether School Environmental Quality is a significant factor influencing the higher preference for private school admissions compared to government schools. The study is driven by the hypothesis that superior environmental quality in private schools contributes to better student health outcomes and overall well-being, which in turn influences parental preference for private education over government schooling. Hence the study is trying to examine the current conditions of Environmental Quality factors in government and private schools. The study also highlights the role of school infrastructure and maintenance practices in shaping the learning environment.

Using a mixed-methods approach, the study gathers data through on-site measurements, surveys, and interviews conducted in both private and government schools. Key Indoor Environmental Quality (IEQ) factors—such as thermal comfort, air quality, lighting, cleanliness, and spatial layout—are assessed to compare the conditions in both school types. The perceptions of parents, students, and educators regarding the impact of these environmental factors on student health, comfort, and academic performance are also examined. Preliminary findings of the hypothesis conclude that environmental quality is a significant determinant of school preference. Private schools tend to offer better-controlled environmental conditions, including sufficient natural lighting, ventilation, and cleanliness, which are perceived by parents as contributing to enhanced student health and academic success. These environmental disparities appear to play a crucial role in the growing preference for private school admissions.

Keyword: *School Environmental Quality (SEQ); Indoor Environmental Quality (IEQ); Parental Preferences; School Infrastructure; Government schools; Private schools; Student's Health and well-being.*

IoT for Smart Connected Buildings

Chaitanya Korra*

ABSTRACT

The integration of Internet of Things (IoT) and Building Information Modeling (BIM) is revolutionizing smart homes and buildings, creating innovative and adaptive environments. Leveraging real-time data collection and digital building modeling, IoT and BIM collectively redefine building performance, enabling dynamic energy optimization, predictive maintenance, and enhanced Indoor Environmental Quality (IEQ). The study investigates these technologies through systematic analysis and selected case studies, showcasing their practical applications and outcomes.

Focusing on IEQ, the paper addresses smart building elements such as air quality monitoring, adaptive lighting, and personalized climate control, enhancing occupant health and productivity. It also evaluates the role of IoT-enabled energy systems in reducing resource consumption while integrating renewable energy sources to achieve sustainability goals. The analysis highlights challenges like data security, interoperability, and standards for IoT-BIM integration, proposing pathways to address these issues. Additionally, the paper examines emerging technologies like 5G, edge computing, and AI/ML to improve real-time building management and decision-making processes.

Case studies, including advanced digital twin applications and innovative smart building implementations, illustrate the transformative potential of IoT and BIM in reshaping the built environment. The findings underscore their significance in fostering energy-efficient, health-focused, and sustainable buildings, paving the way for future-ready construction practices and urban planning. This work aligns with the conference's emphasis on healthier built environments, offering insights into practical strategies for advancing building performance and occupant well-being.

Keywords: Internet of Things, Building Information Modeling, Building Management Systems

References:

1. Kim, D., Yoon, Y., Lee, J., Mago, P.J., Lee, K. And Cho, H., 2022. Design and implementation of smart buildings: A review of current research trend. Energies, 15(12), p.4278.
2. Vijayan, D.S., Rose, A.L., Arvindan, S., Revathy, J. And Amuthadevi, C., 2020. Automation systems in smart buildings: a review. Journal of Ambient Intelligence and Humanized Computing, pp.1-13.
3. Metallidou, C.K., Psannis, K.E. and Egyptiadou, E.A., 2020. Energy efficiency in smart buildings: IoT approaches. IEEE Access, 8, pp.63679-63699.

Techniques and Trends in Indoor Pollen Monitoring and Analysis

Sachin Dhawan * [1], **Mukesh Khare**[2], **Dalip Singh Mehta**[3], **Ekta Sharma**[4]
[1] *School of Interdisciplinary Research, Indian Institute of Technology Delhi, 110016, India*
[2] *Department of Civil Engineering, Indian Institute of Technology Delhi, 110016, India*
[3] *Centre for Sensors, Instrumentation and Cyber-physical Systems Engineering (SeNSE), Indian Institute of Technology Delhi; Department of Physics, Indian Institute of Technology Delhi, Delhi, 110016, India*
[4] *School of Mathematics, Physics and Computing, University of Southern Queensland, QLD 4300, AUSTRALIA*

This study examines current techniques and emerging trends in indoor pollen monitoring and analysis, focusing on key challenges in bioaerosol research. Airborne pollen and fungal spores, important constituents of indoor particulate matter, present significant public health considerations, particularly for individuals with allergies. The research systematically reviews both traditional and advanced methods for detecting and characterizing bioaerosols, highlighting the progression from manual sampling techniques to automated, real-time monitoring systems.

Additionally, the study provides an overview of next-generation monitoring technologies, including optical particle counters, fluorescence-based detection systems, holographic imaging, and conventional volumetric sampling methods. It underscores the increasing integration of artificial intelligence in pollen identification, where machine learning algorithms demonstrate high accuracy when used alongside advanced imaging techniques. These technological developments facilitate automated, high-throughput analysis, thereby reducing manual intervention and processing times.

The study also explores approaches to indoor air monitoring, acknowledging the complexities associated with indoor environments such as intricate airflow patterns, variable pollutant sources, and interactions with other particulate matter. It discusses current methodologies, challenges, and potential future directions. A critical component of the study addresses the importance of standardization, emphasizing the need for consistent terminology, calibration protocols, and quality control measures across the field.

The findings underscore the importance of scalable, real-time bioaerosol monitoring networks to inform evidence-based public health strategies and urban planning. By connecting technological innovations with practical implementation approaches, this study offers valuable insights into next generation monitoring systems capable of addressing the increasing global health impacts associated with pollen exposure.

Keyword: *Bioaerosols, Pollen monitoring, Indoor air quality*

Exploring Emissions from Traditional Cookstoves in Rural Rajasthan: Challenges and Insights

Abhijeet Mondal *, **Anjali Rathee, Dr. Rajyalakshmi Garaga**
Center for Emerging Technologies for Sustainable Development, Indian Institute of Technology Jodhpur, Rajasthan, India – 342037

ABSTRACT

Indoor air pollution caused by biomass-fuelled traditional cookstoves (Chulha's) represents rural India's environmental and public health challenge. This study quantifies particulate matter (PM) emissions and indoor air quality across three cookstove types—mud, cement, and iron—in 100 households from two villages in Rajasthan: Ghadav (60 households) and Jheepasani (40 households). Real-time monitoring of PM1, PM2.5, and PM10, along with carbon dioxide (CO_2) concentrations, was conducted using a calibrated low-cost sensor. Results indicate mud chulhas exhibit the highest emissions, with average PM1 concentrations of 46.4–39.73 $\mu g/m^3$, PM2.5 concentrations of 71.58–62.13 $\mu g/m^3$, and PM10 concentrations of 82.2–70.6 $\mu g/m^3$ in Jheepasani and Ghadav, respectively. Cement chulhas recorded intermediate emission levels (PM1: 70.25–38.5 $\mu g/m^3$; PM2.5: 106.75–59.92 $\mu g/m^3$; PM10: 123.38–68.67 $\mu g/m^3$), while iron chulhas showed lower but hazardous PM concentrations exceeding WHO guidelines. CO_2 levels ranged from 486–659 ppm in Jheepasani and 582–757 ppm in Ghadav, emphasizing variations in ventilation and cooking practices. Structural analysis revealed that Jheepasani's predominance of kutcha houses and limited ventilation exacerbate indoor pollution. Conversely, Ghadav benefited from superior housing designs and ventilation, reducing pollutant levels marginally. Proximity to highways (2.6 km in Jheepasani, 3.1 km in Ghadav) was determined to significantly impact indoor air quality, resulting in exceeding PM concentrations. A preliminary health survey identified 60–70% of women exposed to chulha smoke reporting symptoms such as asthma, eye irritation, sore throat, etc. The findings provide robust evidence to inform rural indoor air quality policies, emphasizing health risk mitigation and environmental sustainability.

Keyword: *Biomass Cookstoves, Indoor Air Pollution, Particulate Matter*

A Study on the Relationship Between the Objective Assessment of the Provisional "CASBEE-Wellness Real Estate" System and Subjective Evaluations from a Questionnaire Survey

Koga Nakamura * [1], **Tatsuya Hayashi**[2]
[1] *Faculty of Engineering, Chiba University, Chiba, Japan*
[2] *Graduate School of Engineering, Chiba University, Chiba, Japan*

This study examines the relationship between objective evaluations conducted using CASBEE Real Estate (Wellness Office Version), pilot version, and subjective evaluations obtained through tenant surveys. A total of 142 properties were investigated, with 135 included in the final analysis. In the objective assessment, only one property received an S rank, while four properties achieved an A rank, indicating that high-grade properties are scarce. Notably, 83 properties, accounting for over 60%, were rated as B-. A moderate correlation was observed between construction year and score, suggesting that newer buildings tend to achieve higher scores. Additionally, a positive correlation was identified between total floor area and score, with larger properties generally scoring higher. For subjective evaluations, data were collected through a corporate entity responsible for building ownership and management. The results showed generally high scores, possibly reflecting response bias due to the data collection method. This could explain the weak correlation found between objective evaluations and subjective responses, highlighting potential gaps between environmental performance metrics and user perceptions. A key finding was that satisfaction levels were highly dispersed among properties rated B-, while dissatisfaction was significantly reduced in properties rated B+ or higher. These findings suggest that office environments may play a more significant role in preventing dissatisfaction than in increasing satisfaction.

References:

Ministry of Land, Infrastructure, Transport and Tourism (MLIT). "Final Summary of the Study Group for Promoting the Spread of ESG Investment." March 28, 2018.
Ministry of Land, Infrastructure, Transport and Tourism (MLIT). "Interim Summary of the Study Group on ESG Real Estate Investment." July 3, 2019.
Hiyama, Kyosuke, Hayashi, Tatsuya, and Kubo, Ryutaro. "Trend Analysis of CASBEE Wellness Office Certified Properties." Japan Architectural Review, vol. 6, no. 3, p. e12403, September 2023.

Keyword: *health and wellness, rating system, workplace productivity*

Developing Residential Building Archetypes for India

Charles Divyateja Bontha * [1], **Aakash C. Rai**[2,3]
[1] *Department of Sustainable Energy Engineering, Indian Institute of Technology (IIT), Kanpur, Uttar Pradesh, 208016, India*
[2] *Department of Sustainable Energy Engineering, Kotak School of Sustainability, Indian Institute of Technology (IIT) Kanpur, Uttar Pradesh, 208016, India*
[3] *Chandrakanta Kesavan Center for Energy Policy and Climate Solutions, Indian Institute of Technology (IIT) Kanpur, Uttar Pradesh, 208016, India*

India's residential building sector (about 300 million homes) consumes about one-fourth of the nation's electricity, which accounts for about 14% of the country's energy-related CO_2 emissions. Thus, decarbonizing India's residential building stock is essential to meet its commitment to achieve net-zero emissions by 2070 and combat global warming. To aid in India's decarbonization efforts, we developed residential building archetypes (an archetype is a model of a building that represents a group of buildings with identical features) that can represent the entire country's building stock and help evaluate the impact of residential energy efficiency programs. To develop those archetypes, we first selected key features (e.g., number of occupants, floor area, appliance ownership, etc.) that impact residential electricity consumption based on a regression analysis. Next, the k-prototype clustering algorithm was used to group residences of similar characteristics together and develop a suitable number of archetypes while striking a balance between representativeness and the number of archetypes. Finally, detailed validation was performed by comparing the state-level energy consumption estimated from the developed archetypes with the actual data. The developed archetypes help gain deeper insights into the residential sector's energy consumption and aid energy simulation and policy efforts at the building stock level.

Keyword: *Decarbonizing homes, Clustering, K-prototype algorithm*

INFLUENCE OF INDOOR ENVIRONMENTAL CONDITIONS ON SLEEP QUALITY AMONG UNIVERSITY STUDENTS IN INDIA

Asmita Addya * [1], **Anubha Goel**[1], **Pawel Wargocki**[2]
[1] *Indian Institute of Technology Kanpur, India*
[2] *Technical University of Denmark, Copenhagen, Denmark*

Impairment of normal sleep structure has been shown to affect cognitive function in both adolescents and adults detrimentally. The interaction between sleep and bedroom environments is sparse, especially in India. This study bridges this gap by examining the influence of indoor environmental conditions on sleep quality in student accommodations. Real-time monitoring was conducted over four nights successively in 139 bedrooms. We collected sleep data via actigraphy watches, Groninger Sleep Quality Survey (GSQS), and sleep diaries. The measured environmental parameters included— temperature, relative humidity (RH), CO_2, PM_{10}, and $PM_{2.5}$. The study highlights the significant impacts of gender and bedroom environment on sleep quality. Male participants had a higher GSQS score (4.4±3.6) compared to females (3.5±2.7), $p<0.05$. Females achieved 25.6% more rapid eye movement (REM) and 20.5% deep sleep (DS) compared to males, though they fell below the recommendations of the National Sleep Foundation. Open windows and doors significantly improved REM and DS while reducing GSQS scores, emphasizing the importance of ventilation in enhancing sleep quality and satisfaction. CO_2 levels were notably lower in open windows and doors compared to closed conditions (609±108 vs. 792±164 ppm, $p<0.001$). Nighttime CO_2, RH, $PM_{2.5}$, and PM_{10} levels were negatively correlated with REM and DS ($p<0.05$), emphasizing the critical role of indoor environments in regulating sleep architecture. These findings underscore the significant influence of the bedroom environment and ventilation on sleep quality. Optimizing indoor conditions can make sleep-conducive bedrooms that enhance both subjective satisfaction and objective sleep indices, supporting healthier living and academic performance.

Keyword: *Sleep quality, GSQS, Indoor environmental quality*

Modeling Indoor Air Quality in a Hostel Mess Using CONTAM

*Abdullahi Mohamed Sheikh Ali[1], Mukesh Khare[2], Arvind K. Nema[3], Sri Harsha Kota[4]**
[1]Department of Civil Engineering, Indian Institute of Technology, Delhi, Hauz Khas, New Delhi, 110016
[2]Professor Emeritus, Department of Civil Engineering, IIT Delhi
[3]Provost, IIT Delhi – Abu Dhabi, Professor, Department of Civil Engineering, IIT Delhi
[4]Associate Professor, Department of Civil Engineering, IIT Delhi
**Correspondence: harshakota@civil.iitd.ac.in*

ABSTRACT

Indoor air quality (IAQ) is crucial for health and comfort, particularly in high-occupancy environments such as hostel kitchens and dining halls, where cooking and human activities significantly contribute to air pollution. This study evaluates IAQ in the kitchen and dining area of Dronagiri Hostel, IIT Delhi, using CONTAM software to model concentrations of carbon dioxide (CO_2), $PM_{2.5}$, and PM_{10} Simulations, corroborated by on-site measurements, indicated that CO_2 concentrations reached a maximum of 1024 ppm in the dining area, slightly above ASHRAE's suggested threshold of 1000 ppm. In the kitchen, PM_{10} levels reached 234 µg/m3, exceeding the WHO recommendation of 45 µg/m^3, while $PM_{2.5}$ also surpassed the WHO permitted 15µg/m^3 limit during cooking hours.

The analysis revealed insufficient ventilation, especially in the kitchen, where non-operational exhaust systems worsened air quality. The findings underscore the need for specific indoor air quality enhancements, such as repairing ventilation systems, optimizing circulation, and integrating advanced purification technology. These approaches could significantly reduce pollution levels, ensuring a better indoor environment for staff and residents. The research highlights the importance of proactive air quality management in high-density residential areas and recommends enhancing existing standards by integrating advanced air purification systems like CAB (Clean Air Bubbles) technology to reduce pollution levels significantly.

KEYWORDS: Indoor air quality, CONTAM simulation, carbon dioxide, particulate matter, hostel kitchen, and dining.

Impact of urban settings on indoor particle number concentrations: A case study from Delhi

MONIKA SHARMA * [1], **Mukesh Khare**[2], **Rajeev Kumar Mishra**[1]
[1] *Department of Environmental Engineering, Delhi Technological University, Delhi 110042, India*
[2] *Department of Civil Engineering, Indian Institute of Technology, Delhi, New Delhi 110016, India*

Indoor air quality is a serious concern because people spend most of their time there. This study aims to estimate the indoor particle number concentrations in Delhi's households (Kitchen and Bedroom) in different settings, i.e., industrial, residential, commercial, and mixed. This study revealed that the quasi-ultrafine particle concentration was found to be high followed by sub-fine and fine particles in all the selected households. The urban indoor region has a quasi-ultrafine particle concentration load of ~10^5-10^6 #cm^{-3} in the kitchen and bedroom microenvironments. The number concentration of coarse particles was found to be lower in all the indoor environments (< 1000 #cm^{-3}). The diurnal concentration of indoor pollutants coincided with the cooking pattern. Apart from urban settings, household infrastructure also played a dominant role in influencing the indoor concentrations of kitchens and bedrooms. The bedroom's particle number concentrations were strongly influenced by cooking activities in the kitchen. Indoor particle number concentration was found to be high in household located in industrial areas, followed by mixed, commercial, and residential zones. The outcomes of the study showed that emissions emanating from varying land-use patterns sources influence outdoor as well as indoor air quality. Urban planning also plays a crucial role in managing air pollution and appropriate strategies and control measures should be initiated by urban local bodies and policymakers.

Keyword: *Indoor air quality, PNC, Households, Microenvironments, Urban settings*

MEASUREMENT OF HYGROTHERMAL PERFORMANCE OF AN EVAPORATIVELY COOLED, DOUBLE-WALLED SYSTEM WITH CEMENT AND LIME MORTAR.

Alankrit *[1], **Rashmin Mohan Damle**[2]
[1] *Masters' in Building Energy Performance, CEPT University, Ahmedabad, India*
[2] *Faculty of Technology, CEPT University, Ahmedabad, India*

This ongoing research evaluates the hygrothermal performance of an evaporatively cooled, double-walled enclosure using lime and cement mortar. Inspired by the pot-in-pot clay refrigerator, the study adapts this passive cooling method for buildings. The traditional system, where a smaller pot sits inside a larger pot with water-saturated sand in between, enables evaporative cooling. Studies show such refrigerators extend shelf life by 50% [1] and save 40% energy than conventional refrigeration[2]. The research aims to extend the viability of evaporative cooling to building components by developing a double-walled system. The hygrothermal analysis of lime and cement mortar evaluates how differences in porosity and moisture retention impact cooling effectiveness, indoor temperature reduction, and energy savings.Two identical cells (0.6 m^3) have been built one with lime mortar and the other with cement mortar in Ahmedabad's hot, dry climate. The cavity between walls is filled with water-saturated sand for evaporative cooling.Indoor temperature, surface temperature, and relative humidity are to be monitored from January 4 to April 20, 2025. Between January 4th to January 30th, the lime cell reduced indoor temperature by 3°C below ambient, compared to 2.6°C in the cement cell. The average surface temperature difference between interior and exterior was 2°C in the lime cell and 0.57°C in the cement cell. The cement cell maintained 8% lower indoor relative humidity than ambient, while the lime cell achieved 6.3%. The lime cell provided more balanced moisture regulation, proving more effective for passive cooling. Further insights during summer months are expected as the study progresses.

References:

[1] R. Choudhary et al., "Evaporative Cooling in Traditional Clay Pot Refrigeration," Journal of Sustainable Engineering, vol. 12, no. 4, pp. 215-225, 2023.
[2] A. Ghosal and S. Rath, "Energy Savings and Efficiency in Passive Cooling Systems," Energy & Buildings, vol. 198, pp. 345-355, 2019.

Keyword: *Evaporative cooling, Hygrothermal performance, Passive cooling, Lime mortar, Sustainable construction*

Impact of Sun Path on Thermal Profile of Indoor Environment in School Classrooms in Malaysia

Yassir Arafat[1], Sheikh Ahmad Zaki[1]*, Noor Alam[1], Naja Aqilah[1], Fitri Yakub[1], Teh Khar San[2], Yam Kah Wei[2].

[1]Department of Mechanical Precision Engineering, Malaysia-Japan International Institute of Technology, Universiti Teknologi Malaysia, Kuala Lumpur 54100, Malaysia
[2]San Miguel Yamamura Woven Products Sdn. Bhd., Lot 9 & 10, Jalan Usaha 4, Kawasan Perindustrian Ayer Keroh, 75450 Ayer Keroh, Melaka

*Email: sheikh.kl@utm.my

ABSTRACT

The thermal performance of building envelopes, particularly roofs, is crucial for indoor thermal comfort in tropical climates like Malaysia. Roofs are the main pathway for solar heat gain, and elevated indoor temperatures can negatively impact student comfort and learning outcomes. The study aims to provide in-situ measurements of the thermal profile across roof assemblies in two secondary schools, capturing temperature variation at multiple layers and quantifying heat transfer into classrooms. Several temperature measuring sensors are placed from 0.5 m above the floor to the rooftop to record daytime data. Understanding vertical temperature gradients is crucial for improving indoor environmental conditions and reducing the amount of heat that is gained in classrooms by improving tactics regarding radiant barriers, roof insulation, and attic ventilation.

Keywords: indoor temperatures, Vertical Thermal Gradient, school classrooms

1. Introduction

The thermal performance of the building envelope, especially roofs, plays a crucial role in ensuring indoor thermal comfort in naturally ventilated classrooms in tropical climates like Malaysia. Roofs serve as the main pathway for solar heat gain, and elevated indoor temperatures may adversely impact on student comfort and learning outcomes. Researchers found a significant relationship between the physical environment and school productivity. Environmental comfort is one of the key elements that has an impact on academic achievement [1]. In recent research, researchers have mentioned that students' performance increases with thermal perception and classroom conditions [2]. More precisely, thermal comfort in a classroom significantly impacts students as they spend more than half of the day in the school. Thermal comfort is defined as a state of mind regarding satisfaction, varied with personal and ambient parameters of the thermal environment [3]. Several researchers have examined the effect of thermal comfort on room temperature. A specific temperature cycle between 25 and 28°C can create thermal comfort by reducing energy consumption [4]. Temperature variation on different scales can be a reason for an unstable thermal environment in the classroom, which increases thermal discomfort. Moreover, the classroom temperature also does not suddenly increase drastically; it is changing over different times of day. Researchers found that the transmission of solar radiation indoors significantly impacts the thermal comfort of individuals near windows [11]. It is imaginable how much the temperature influences the classroom environment without any built-in thermal insulation facilities. Since the thermal performance of a building is highly impacted by the solar absorbance of the roof, it is necessary to consider what type of roof materials are used for the building. The researcher mentioned that between 20% and 90% of radiation is absorbed in the roof surface, and it can reach up to 1 kW/m^2 during clear sky [12]. Metallic roof tile is used as roofing material in those schools. The metallic roof exhibited the highest outdoor temperature when compared to ceramic and concrete roofs [13]. The maximum indoor temperature was recorded on the metallic roof, and the highest heat flux observed among the three types of roofs was also on the metallic roof.

According to Malaysian Meteorological Department (MET) Malaysia, the outdoor temperatures in May 2025 are between 25.4 and 33.4°C, on average, while the weather is tropical, which is mostly humid and hot. The researcher claimed that roof material and colour directly impact the solar heat gains, which affect the indoor thermal comfort conditions of public school buildings in Malaysia [5]. To improve indoor environmental quality and optimize HVAC systems, it is crucial to know the vertical temperature gradients, as they readily impact occupants' comfort.

This study aims to provide in-situ measurements of the thermal profile across roof assemblies in two secondary schools, capturing temperature variation at multiple layers and quantifying heat transfer into classrooms. The vertical temperature distribution in a classroom starts from 0.5 m above the floor to the rooftop. Although according to ASHRAE Standard 55, it is required to start measuring the room temperature from 1.0 m above the ground, in this study, it is considered 0.5 m above the ground floor.

2. Methodology

2.1 Location and climate

The study was conducted in three classrooms, each at two secondary schools, the classroom temperature and vertical thermal gradient were analyzed for SMK Taman Seri Rampai (3°11'31.2"N 101°43'57.9"E) and SMK Padang Tembak (3°10'48.5"N 101°43'24.8"E), Kuala Lumpur, Malaysia. Average toasty temperature feels stuffier due to its location in a hilly equatorial area. Field measurements were carried out from May 15 to July 11, 2025, on the top floor (4th) of both the school buildings, which are situated facing the south direction. The sun's path significantly impacts both schools' indoor temperatures. Solar radiation plays a crucial role in shaping indoor temperatures, primarily by enhancing heat gain through building envelopes, particularly in regions that experience considerable sunlight exposure [6].

Fig.1 (a and b) helps to assume how directly the classroom is heated during the day. Moreover, data from MET Malaysia confirms that the highest temperatures this year were recorded from March to May. The effects of solar incidence on indoor environments and their occupants are substantial, influencing indoor thermal comfort and potentially causing unwanted overheating throughout the year [7]. In Recent years, the mean air temperature and relative humidity in Kuala Lumpur were 27.7°C and 85%, respectively [8].

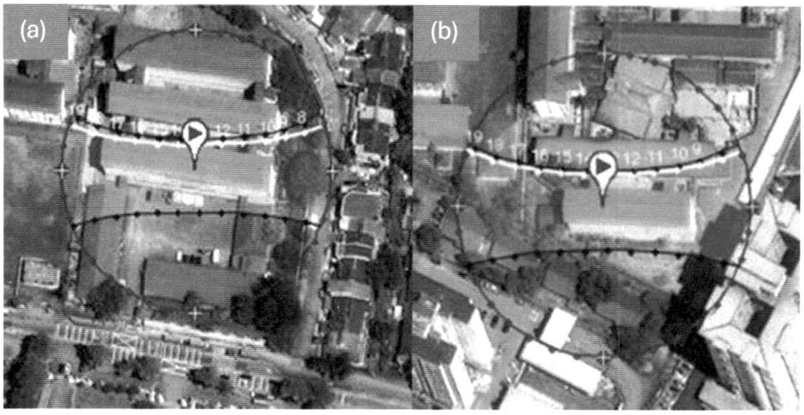

Fig. 1. Sun path above (a) SMK Taman Seri Rampai (b) SMK Padang Tembak.

2.2 Selected schools and classrooms

Large Standard size of classroom in Malaysia, size = Height (H) × Length (L) × Width (W) = 3.60 m × 9.23 m × 7.31 m = 242.89 [9]. The Public Works Department (PWD) decides on the building materials and specifications for public school buildings, which are followed by the contractor, where mostly common building materials are used without considering any thermal insulation properties. The thermal conductivity of common building insulation materials increased approximately linearly with increasing temperature, which is characteristic of materials and varies depending on relative humidity [10].

2.3 Field measurement procedure

For the measurement purpose, a top-floor classroom was selected for both schools, which is in the middle position of the building. Public schools in Malaysia are equipped with louvred windows that were open during measurements to allow natural ventilation. Although the classroom had three ceiling fans in the middle and two wall fans on the left side, those were only used during class hours. The indoor globe and radiant shielded temperature are measured by the HOBO U10 data recorder as shown in Fig.2. Measurement was conducted in three separate time frames, daytime, nighttime, and the whole day, to collect sufficient data.

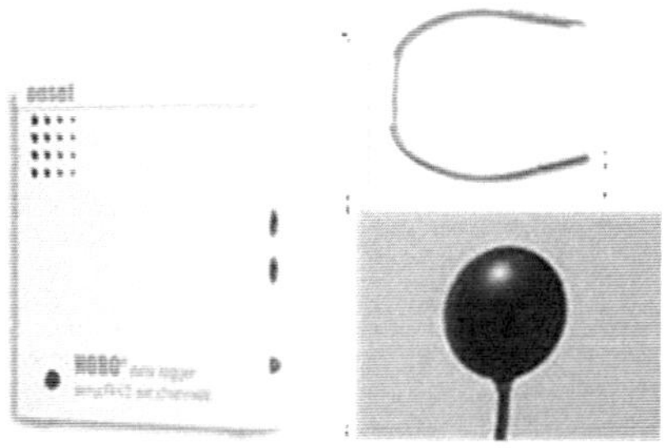

Fig. 2. Hobo indoor temperature recorder U12-013.

The indoor air temperature (T_i) and relative humidity (RH_i) were measured at three heights (0.5 m, 1.5 m, and 2.5 m) to monitor thermal stratification. Moreover, thermocouples are installed in the attic air space, inner ceiling surface, outer roof surface, and inner roof surface to measure temperature variation.

Fig. 3. Indoor temperature measurement.

Besides, to measure outdoor temperature (T_O), two thermocouples were installed outside the classroom. Fig.3 illustrates the position of Hobo for indoor temperature at different heights. Thermocouple is a type of temperature sensor; its low cost and self-powered properties make it preferable for having a stable reference temperature. It consists of two different thermoelements that allow the generation of a potential difference or voltage based on the thermoelectric phenomenon [14]. Heat Flux Transducer: Installed on the ceiling surface to measure heat flow into the classroom. All equipment is set up for data collection in a 5-minute interval, and all thermocouples and heat flux sensors were connected to a Graphtec data logger GL820 Fig.4. HOBO sensors for indoor and globe temperatures are recorded at similar time intervals. The position of sensors above the ceiling is illustrated in Fig.4.

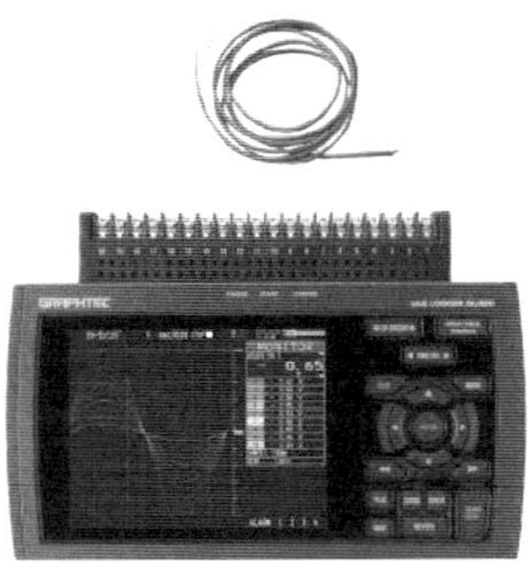

Fig. 4. Data logger GL820

3. Results and Discussion

Fig.5(a). shows that for SMK Sri Rampai, room temperature was almost the same for three different heights, which start below 30°C, and except for a few hours of daytime, any drastic change is not visible. Rooftop temperature shows fluctuation starting near 8:00, and it peaks above 65°C after 14:00, then a sudden fall in temperature below 30°C is noticed around 15:00. Although a temperature rise happens at 4:00 pm, it gradually decreases. After 6:00 pm, the temperature from the ceiling to the room decreased, but the room temperature increased. Fig.5(b) SMK Padang Tembak also illustrates similar results with small variations, where roof temperature crossed over 70°C, and above the ceiling, more fluctuation is visible during the daytime. From both graphs, it is common for the rooftop temperature to decrease at night to below 25°C, which is the lowest in the graph. Still, during the daytime, it reaches almost double the average daily temperature in Malaysia.

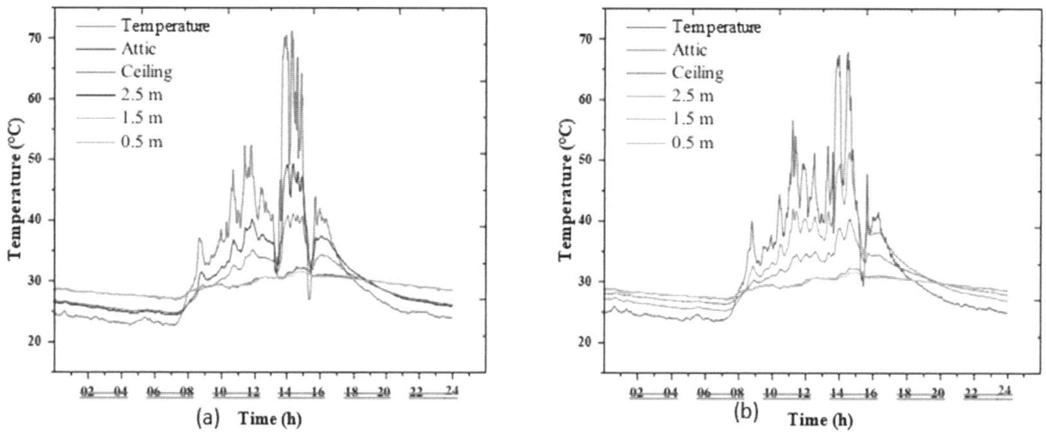

Fig. 5. Hourly average temperature (a) SMK Taman Seri Rampai (b) Roof SMK Padang Tembak

3.1 Comparison of vertical gradient temperatures

In terms of natural ventilation, decreasing solar heat absorption in the roof could contribute to moderate fluctuations in indoor temperature [15]. The vertical gradient of both schools is calculated from the temperature difference between the ceiling and the roof. The vertical gradient (∇T),

$\nabla T = \Delta T / \Delta Z$

ΔT = Temperature difference between two points (°C). Δz = Vertical distance between the two points (m).

Vertical profiles of the air temperature from floor to rooftop of the classroom are shown in Fig.
6. The present values are the relative temperatures concerning the floor of the classroom, time average of the entire period, including the daytime (7:00-19:00) and nighttime (19:00-7:00). The temperature on the inner surface of the ceiling board exhibited variations between nighttime and daytime conditions. Throughout the daylight hours, the temperature demonstrated a

notable rise with increasing elevation, reaching its maximum at the inner surface of the roof tile. The temperature differential between the attic and the inner roof surface was notably substantial, attributed to the heating effects of solar radiation. The roof tile's outer surface temperature was lower than that of the inner surface. This discrepancy is likely attributed to transient evaporation cooling occurring during showers, as well as the heat storage properties of the tile. On the other hand, the temperature above the ceiling during the nighttime exhibited a considerable decrease along the height direction, probably owing to the cooling of the nocturnal radiation. The room air temperature is 6°C and 8°C higher than the outer roof surface temperature during the nighttime for SMK Taman Seri Rampai (SMK-SR) and SMK Padang Tembak (SMK-PT). As a result, SMK PT is much hotter than SMK SR, even in the early morning.

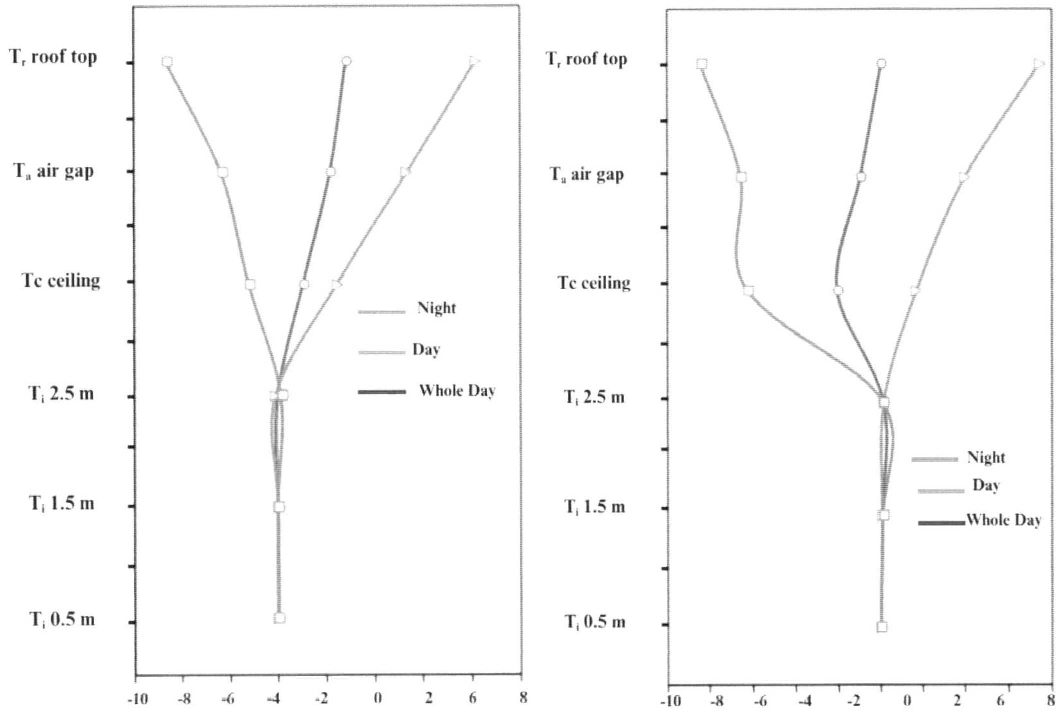

(a) (b)

Fig. 6. hourly average temperature (a) SMK Taman Seri Rampai (b) SMK Padang Tembak

3.2 Indoor temperature

Fig.7(a). The graph shows the indoor temperature variations of three classrooms (Room 1, Room 2, and Room 3) in SMK-SR. The temperature fluctuates between 27°C and 36°C in all rooms, with Room 3 (green line) generally experiencing higher temperatures than the other two. All rooms follow a similar periodic pattern, likely influenced by daily factors such as sunlight, air conditioning, or ventilation, with temperatures peaking during the day and dropping at night. Room 3 exhibits the largest fluctuations, reaching 36°C, while Room 2 (orange line) stays lower, ranging from 28°C to 34°C. Room 1 (blue line) falls between the other two, fluctuating between 29°C and 34°C. These findings suggest that while all rooms experience similar temperature cycles, Room 3 tends to be the warmest, indicating that improvements in ventilation or cooling systems might be necessary to enhance comfort in this room. On the other hand, Graph Fig.7(b). illustrates the temperature fluctuations in SMK-PT classrooms. Room 1 (blue line) shows moderate temperature changes, ranging from 28°C to 35°C, with a consistent periodic rise and fall, and slight increases towards the middle of the observed period. Room 2 (orange line) experiences more erratic fluctuations, with temperatures ranging from 29°C to 34°C, showing sharper peaks around mid-June and late June. This suggests varying external conditions or HVAC inconsistencies. Room 3 (green line) displays the widest temperature range, from 27°C to 35°C, with dramatic peaks and dips, especially between mid-June and early July. The temperature variations in all rooms peak during the day, likely due to external factors such as sunlight or internal heat from activities or equipment. Overall, the temperatures in the rooms appear to rise during late June and early July, possibly due to warmer outside weather or changes in the classroom climate.

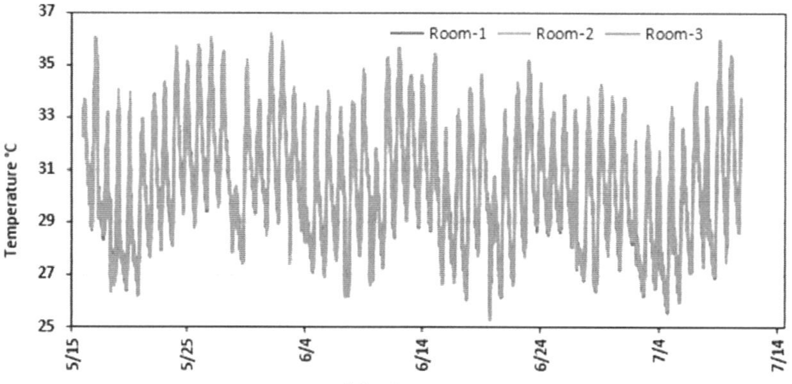

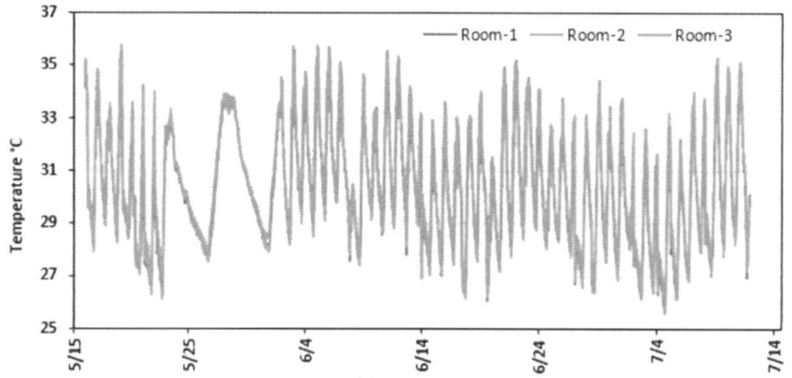

Fig. 7. Variation of indoor Temperature (a) SMK-SR, (b) SMK-PT

Table 1 presents indoor temperature statistics for school classrooms in Kuala Lumpur, specifically comparing data from SMK Taman Seri Rampai and SMK Padang Tembak. Across the three classrooms at SMK Taman Seri Rampai, maximum temperatures ranged from 35.6°C to 35.8°C, while minimum temperatures varied between 25.6°C and 26°C. The mean temperatures were relatively consistent, ranging from 30.2°C to 30.6°C, with standard deviations (SD) around 2.1°C to 2.2°C, and coefficients of variation (CV) between 7.0% and 7.3%, indicating moderate variation in indoor temperatures. At SMK Padang Tembak, although all entries are labeled "Classroom-1," the readings suggest measurements from different rooms or periods. Maximum temperatures reached as high as 36.2°C, with minimums as low as 25.3°C, and mean temperatures consistently around 30.2°C to 30.2°C. Standard deviations here ranged from 2.2°C to 2.3°C, and CV values were slightly higher, between 7.40% and 7.54%, indicating slightly more temperature variability than SMK Taman Seri Rampai. Overall, both schools experienced similar thermal conditions, with minor differences in temperature stability.

Table. 1 Indoor Temperature Statistics for School Classrooms in Kuala Lumpur.

SMK Taman Seri Rampai					
Room	Max Temp (°C)	Min Temp (°C)	Mean Temp (°C)	*SD* (°C)	*CV* (%)
Classroom-1	35.7	25.6	30.6	2.2	7.3
Classroom-2	35.6	25.6	30.2	2.2	7.3
Classroom-3	35.8	26.0	30.5	2.1	7.0
SMK Padang Tembak					
Classroom-1	35.9	25.5	30.2	2.2	7.4
Classroom-1	36.2	25.4	30.2	2.3	7.5
Classroom-1	36.0	25.3	30.2	2.3	7.5

Note: Max Temp: maximum temperature; Min Temp: minimum temperature; *SD:* standard deviation; *CV*: coefficient of variance

4. Conclusion

This research shows extensive temporal and geographic temperature data gathered from the roof assemblies of classrooms in Malaysian secondary schools. Although the result shows that classroom of SMK Padang Tembak was warmer than SMK Taman Seri Rampai but overall, the data was acquired from the roof assemblies of classrooms suggest that there are significant temperature differences between the metal deck layers of the roof, as well as a delay in the flow of heat into the classrooms, which contributes to an increased level of thermal discomfort. Variations in attic ventilation likely have a substantial impact on the thermal performance of the building. To reduce the amount of heat gained in classrooms, the statistics suggest a need for improved tactics regarding radiant barrier, roof insulation, and attic ventilation. In order to provide a more comprehensive knowledge of seasonal effects and to inform retrofit recommendations targeted at improving energy efficiency and thermal comfort in school buildings, additional analysis will be conducted through increased data collection.

Acknowledgement

The Universiti Teknologi Malaysia financially supported this work through the F4+ High

Tech Grant [Vot 00Q82]. This research was funded by a grant from the Ministry of Higher Education of Malaysia (FRGS Grant R.J130000.7824.4X172).

References

[1] Wong, L. T., Mui, K. W., and Tsang, T. W. "An Open Acceptance Model for Indoor Environmental Quality (IEQ)." Build. Environ. 142, (2018): 371–378.

[2] H. Azemati, Z. Aminifar, and S. Pourbagher, "Effective environmental factors on designing productive learning environments," Armanshahr Architecture & Urban Development, 11(22), 1-8, Spring 2018.

[3] C. Z. Y. Chong Zi Yao, M. N. A. Nor Azl, A. Hariri, A. A. Muhamad Damanhuri, and M. S. S. Mustafa, "Preliminary study on student's performance and thermal comfort in classroom," Journal of Advanced Research in Fluid Mechanics and Thermal Sciences, vol. 101, no. 1, pp. 59-72, 2023.

[4] S. K. Sansaniwal, J. Mathur, and S. Mathur, "Review of practices for human thermal comfort in buildings: present and future perspectives," International Journal of Ambient Energy, vol. 43, no. 1, pp. 2097-2123, 2022.

[5] S. Li, X. Jia, C. Peng, Y. Zhu, and B. Cao, "Effects of temperature cycles on human thermal comfort in built environment under summer conditions," Science of The Total Environment, vol. 912, p. 168756, 2024.

[6] N. M. Alwi et al., "Retrofitting measures for climate resilience: Enhancing the solar performance of Malaysian school buildings with passive design concepts," in IOP Conference Series: Earth and Environmental Science, 2022, vol. 1102, no. 1: IOP Publishing, p. 012014.

[7] Antonik-Popiołek, P. (2021). Influence of solar radiation to the temperature inside a three-layer partition in winter season. Acta Scientiarum Polonorum Architectura, 20(2), 75–82. https://doi.org/10.22630/aspa.2021.20.2.16

[8] N. S. Firman, S. A. Zaki, N. W. Tuck, M. K. Singh, and H. B. Rijal, "A study on adaptive thermal comfort and ventilation in Malaysia secondary school classrooms of tropical climate," Building and Environment, p. 112701, 2025.

[9] N. Norazman, A. I. Che Ani, W. N. W. Ismail, A. H. Hussain, and K. N. Abdul Maulud, "Indoor environmental quality towards classrooms' comforts level: Case study at Malaysian secondary school building," Applied Sciences, vol. 11, no. 13, p. 5866, 2021.

[10] Y. Wang, S. Zhang, D. Wang, and Y. Liu, "Experimental study on the influence of temperature and humidity on the thermal conductivity of building insulation materials," Energy and Built Environment, vol. 4, no. 4, pp. 386-398, 2023.

[11] Hwang, R., Fang, P., & Chen, W. (2023). Impact of solar radiation on indoor thermal comfort near highly glazed façades in a hot-humid subtropical climate: An experimental evaluation. Building and Environment, 243, 110725.

[12] A. M. Al Yacouby, M. F. Khamidi, M. F. Nuruddi, S. A. Farhan, and A. E. Razali, "Study on

roof tile's colors in Malaysia for development of new anti-warming roof tiles with higher Solar Reflectance Index (SRI)," in 2011 National Postgraduate Conference, 2011: IEEE, pp. 1-6.

[13] Nur'aini, R. D. (2014). Investigation of Ceramic Roof, Metallic Roof and Concrete Roof Thermal Performance in Tropical Climate. NALARs, 13(2).

[14] B. Arman Kuzubasoglu and S. Kursun Bahadir, "Flexible temperature sensors: A review," Sensors and Actuators A: Physical, vol. 315, p. 112282, 2020.

[15] Tuck, N. W., Zaki, S. A., Hagishima, A., Rijal, H. B., & Yakub, F. (2020). Affordable retrofitting methods to achieve thermal comfort for a terrace house in Malaysia with a hot–humid climate. Energy and Buildings, 223, 110072. https://doi.org/10.1016/j.enbuild.2020.110072

ACOUSTIC COMFORT BENEFIT ANALYSIS OF SOUND ABSORBING MATERIALS: A CASE STUDY

Mahesh K. * , **Shailee Achal Goswami, Sreeja S.**
Saint Gobain India Private Limited (Research & Development), Chennai, India

Sound-absorbing materials are designed to enhance sound quality by minimizing echoes and reducing unpleasant reverberation. This paper presents a case study on the acoustic comfort benefits of sound-absorbing materials in a corporate office setting, focusing on cafeteria spaces. Initial experimental measurements showed that the cafeterias, treated with solo panels and baffles, had reverberation times within the acceptable limit of 1 second as per the WELL building standard, confirming the effectiveness of the installed acoustic treatments. These results were supported through simulations conducted in Odeon software. Further, to analyse the impact of the sound-absorbing materials, the reverberation times were simulated for the same spaces without acoustic treatments. The results indicated significantly higher reverberation times of 2.8 s and 2.6 s for the two cafeteria spaces under untreated conditions, compared to 0.96 s and 0.98 s in their acoustically treated state, highlighting the poor acoustic comfort in the absence of sound-absorbing materials. Notably, the investigation indicated that over 70% of sound absorption in these areas was attributed to sound-absorbing materials at ceilings. This highlights the critical role of these materials in achieving optimal acoustic comfort and emphasizes their necessity in designing functional and comfortable spaces.

Keyword: *Sound absorption, Reverberation time, Speech transmission index*

Visual and Thermal Comfort: An Exploratory Study on Understanding a Key Interface

Jaydeep Vaghela * , **Krishan Upadhyay**
Institute of Architecture and Planning, Nirma University

Recent study has suggested that indoor thermal condition and lighting levels are correlated for visual comfort in indoor spaces, although the fact that layers of this correlation are unclear. Lighting influences how the occupant perceives and senses a space, while the mind indicates satisfaction with the thermal environment and it is evaluated by subjective evaluation as thermal comfort. This study investigated the research gap by conducting field and experimental study to establish the combined effects of lighting and thermal environment on visual and thermal comfort, ultimately contributing to energy efficiency and improved indoor environmental quality. The major component of the study is understanding how variations in lighting schemes influence thermal comfort perception across different thermal conditions.

The optimum result is determined by proposing an experiment in a controlled climate chamber at IIT Roorkee, where illumination and thermal conditions are varied. Daylight, artificial light, and thermal variables are controlled. Each experiment session lasts 5 - 7 hours, involving participants in the 20-30 year age group. All experimental sessions are conducted exclusively on sunny days, with experiments on cloudy days (3 to 4 days) deliberately excluded to maintain consistent environmental conditions. During the experiment, careful consideration is given to participants' factors, including their clothing, color vision, overall health status. Each participant consistently takes part in all experimental sessions, follows the test configurations, and provides subjective feedback on visual and thermal comfort as well as their overall perception of the indoor environment. The key findings are expected to apply to practical applications in offices.

Keyword: *Visual Comfort, Thermal Comfort, Climate Chamber, Indoor Environmental Quality, Office Environment.*

Determinants of Comfort-Driven Energy Consumption in Indian Dwellings: A Household-level Perspective

<u>KRITHIKA PANICKER</u> * **, Prashant Anand, Abraham George**
Indian Institute of Technology Kharagpur, Kharagpur, India

Rapid urbanization in developing countries like India, is driving a surge in energy demand and associated CO_2 emissions, requiring urgent action. This, coupled with the increasing frequency and severity of extreme weather events in the country, intensifies the need for energy-intensive technologies, particularly in residential buildings, to maintain comfortable indoor environments. The relationship between residential diversity and energy consumption patterns in India is complex and dynamic, since the types of dwellings, socio-economic status, and lifestyle behaviors vary widely across regions. This study explores how household characteristics, behaviors, and lifestyle choices influence residential electricity consumption, with a focus on improving indoor comfort. This research was carried out using questionnaire survey, and semi-structured interviews conducted across 200 households representing diverse socioeconomic contexts and building typologies. The study incorporated user-defined perceptions of a 'comfortable home', obtained through interviews, to investigate the electricity expenditure associated with it. A general linear model was used to identify the key factors affecting household energy consumption. Household profiles were then created through cluster analysis, and each profile was examined for energy-related behavioral patterns. The analysis demonstrates that space-cooling related comfort significantly impacts household electricity consumption; this impact however varies across different household profiles. Comfort-related energy expenditure is also influenced by behavioral adaptations, including mixed-mode operations, adjustments in seasonal clothing, and the utilization of traditional passive practices. A key finding is that interventions like building retrofits, energy-efficient behaviors, and awareness campaigns should be tailored to household profiles, creating a synergy between energy efficiency and indoor comfort through targeted interventions.

Keyword: *Residential Energy Consumption, Occupant Behaviour, Indoor Comfort*

Seasonal Changes of Adaptive Behaviors for Improving Thermal Comfort in Urban Houses- A Case of Kharagpur, India

Subhagata Mukhopadhyay * [1], **Kazuma Kawajiri**[1], **Nikhil Kumar**[2], **Dulis Dulis**[3], **Tetsu Kubota**[1], **Shankha Pratim Bhattacharya**[4], **Hanief Ariefman Sani**[1]

[1] *Graduate School of Advanced Science and Engineering, Higashi Hiroshima, Japan*

[2] *Department of Architecture, Planning and Design, Indian Institute of Technology, (BHU) Varanasi, Uttar Pradesh, India*

[3] *Graduate School of Innovation and Practice for Smart Society, Higashi Hiroshima, Japan*

[4] *Department of Architecture and Regional Planning, Indian Institute of Technology Kharagpur, West Bengal, India*

Indian cities in warm-humid climatic zone experience three seasons, including summers, winters and monsoons. Summers experience high temperatures of around 45°C and low humidity, monsoons bring heavy rainfall with nighttime humidity reaching 100%, and winters are as cold as 12°C and dry. Occupants adapt to seasonal changes through behavioral adjustments, like opening windows, changing clothes and rooms. This study aims to understand these behavioral patterns by long-term environmental measurements and questionnaire-based investigations in three households with three family members each, in Kharagpur, since July 2023. Air temperature (AT), relative humidity (RH), daily activities and comfort levels were monitored for 14 days in each season. Results showed that in monsoon, indoor AT ranged from 27–32°C with RH 80-90%, while outdoor AT was 24–36°C with RH 89-100%. Occupants stayed indoor 13 hours daily, where natural ventilation, frequent bathing were common adaptive behaviors. In winter, the average AT and RH indoors were 23°C and 60%, whereas outdoor were 12°C and 78%. Increased physical activity, extended sleeping hours, and use of shaded spaces, could be seen with 12 hours indoor daily. Summers exhibited different patterns but consistent indoor AT stability (30°C in average), with indoor RH between 70-75% and outdoor between 60-90%. Occupants stayed indoors for 17 hours daily where they used AC, closed windows during the daytime and changed the rooms for activities, as thermal comfort was unassured, due to poor ventilation despite high thermal mass of houses. This study highlights opportunities to enhance thermal comfort by optimizing occupant behavior and passive strategies.

Keyword: *Occupant Behavior, Field Measurement, Warm-Humid Climate*

Post-disaster reconstruction: Thermal and Visual Comfort analysis of reconstructed school buildings after Gorkha Earthquake 2015 (Case of type design of Lalitpur district)

Nira Shrestha * , **Samjhana Bhattrai, Alfa Shrestha**
Tribhuvan University

Several school buildings were rebuilt which were destroyed by the massive earthquake followed by numerous aftershocks of April 2015. The school buildings i.e. type design have mushroomed considering earthquake resistance design. However, the study on its indoor thermal and visual environment is still lacking. The purpose of this research is to study the thermal comfort and visual comfort of the type design constructed at the Lalitpur district using a simulation tool whereas a field study based on survey questionnaires with secondary students and teachers is used to validate the simulated data. The study was carried out in 9 naturally ventilated classrooms of four different schools during the winter season. Besides collecting energy consumption data and environmental data, a self-reported questionnaire and an observation form were employed to record children's thermal comfort and visual comfort. From four schools, 154 students aged 11-16 and 9 teachers were surveyed and questionnaires were collected. The methodology adopts a holistic approach, correlating students' post-occupancy evaluations with software simulations. Various scenarios of naturally ventilated school buildings were simulated. Case A served as the base-case scenario, while other scenarios involved modifications to building orientation, envelope design, glazing materials, and openings. The results indicated that the base case exhibited discomfort hours and insufficient daylight. These issues were mitigated in Case C by introducing optimal passive design parameters, resulting in reduced discomfort hours and improved daylight levels.

Keyword: *Post-disaster Reconstruction, School Type design, Thermal Comfort, Visual Comfort*

Illuminating Design in Indian Context: Exploring the Relationship between WWR, Latitude, Orientation, and Interior Daylighting in Built Spaces

Anil Kumar Kasichainula, Hari Priya Chowdary Annabathuni

First Author: Anil Kumar Kasichainula

ORCID: 0000-0003-3580-1978

Corresponding email ID: anilkasichainula@gmail.com

ICFAI School of Architecture, ICFAI Foundation for Higher Education, Hyderabad, India

Second Author: Hari Priya Chowdary Annabathuni

ORCID: 0009-0000-5696-6689

Environmental Performance and Design Laboratory (EPDL), Faculty of Architecture and Town Planning, Technion – Israel Institute of Technology.

Abstract

To provide a better-built environment, green rating systems and relevant codes set a minimum daylighting standard. Mostly, the set standards remain the same across the country and on a global scale. Literature sources mention that sky illumination varies with latitude. This paper attempts to understand the design implications of following relevant codes as the latitude varies, in terms of window wall ratio (WWR) and orientation for a higher educational building. A parametric model has been created with fenestrations on one side using Rhino-Grasshopper for analyzing annual lighting simulations with Honeybee tools and the Radiance engine. The variables in the model are WWR (10% to 60%) and Orientation (8 directions), for 32 cities in India at various latitudes, leading to 13056 variations. The Colibri plugin has been used to automate the result generation for all 13056 variations. The results indicate that the effect of orientation on internal daylight becomes more profound as latitude increases. For scenarios having fenestrations on Southeast, South, and Southwest, to achieve the target daylighting levels set by Green Ratings, the required sizes of windows decrease as latitude increases, and for other scenarios with fenestrations on Northwest, North, and Northeast, the required sizes of windows increase as latitude increases. This suggests that southern skies become brighter as we go up in latitude, and the light from northern skies decreases as latitude increases. This research suggests reconsidering some finer aspects of daylighting standards in green ratings.

Keywords

Daylighting; Radiance; Indian Codes; Parametric Model

Analysing Thermal Comfort and Energy Performance in Lecture Rooms using Occupant Surveys

Jayati Dudani[1], Vallary Gupta * [1], Chirag Deb[2]
[1] *Indian Institute of Technology Bombay, Mumbai, India*
[2] *The University of Sydney Sydney, Australia*

With buildings accounting for around one-third of the total energy in India and people spending 80-90% of their time indoors, maintaining indoor environment quality (IEQ) while minimizing energy usage is crucial. However, research focusing on simultaneous reduction in energy consumption and enhancing IEQ is scarce, especially in the context of educational buildings. This study aims to initiate a systematic framework for optimizing IEQ and energy consumption in educational settings. It employs both secondary and primary survey-based data to simulate the energy performance of lecture rooms at IIT Bombay, India. Multi-objective optimization is employed to obtain the optimum range for set points and occupancy for enhanced energy performance.

The results show that both occupancy and lighting levels exert considerable effects on energy consumption and IEQ. Indoor air temperatures at around 23-25°C were found suitable for indoor thermal comfort across different occupancy levels. The findings underscore the importance of considering room-specific parameters and cooling system types. This offers valuable insights for building designers and policymakers. This research paves the way for developing a cyber-physical framework that leverages sensors, IoT-based data collection, and advanced learning approaches to transform future educational spaces into energy-efficient and occupant-centric sustainable environments.

Keyword: *Indoor Environment Quality, Energy Consumption, Building Energy Simulation*

Bi-level strategies for Micro-Zonal Occupant-Centric Control (MZOCC) to reduce HVAC energy consumption

Jeslu Celine Jacob * , **Debapratim Pandit, Joy Sen**
Indian Institute of Technology (IIT) Kharagpur, India

Micro-Zonal Occupant-Centric Control (MZOCC) is a novel strategy to selectively air-condition occupied virtual micro-zones within large rooms. Unlike most personal environmental modules (PEMs), which require additional infrastructure and high capital investments, MZOCC upgrades standard HVAC systems (e.g. centralized air-conditioning with overhead-diffusers) for localised air-conditioning. Previous publications highlight the challenges in containing cool air in small regions using MZOCC. This study develops a systematic method to address these challenges and implement MZOCC. The framework consists of: **(1) Micro-Zoning (MZ),** which divides the thermal zone based on activities and HVAC systems, and **(2) Occupant-Centric Control (OCC)**, which optimizes independent operation of diffusers based on occupancy and comfort preferences.

The study proposes **bi-level optimization strategies**, developed from a survey of forty Indian offices and experimentally validated CFD-based simulations in six offices. Weekly office usage of thirty occupants is also analysed. The potential of MZOCC is demonstrated through a simulation-based case-study using an experimentally validated CFD model of an open-plan office in Hyderabad, which accommodates 60 occupants in 693 sqm.

Results of the case-study show that MZOCC saves about 60% HVAC energy under randomized occupancy and 80% on grouping occupants based on schedules and thermal preferences. Implementing MZOCC in the case-study office requires an additional investment of ₹3,09,207 ($3,599.09) for occupancy detection and automatic controllers. However, MZOCC reduces daily cooling costs from ₹11,882 ($138.30) to ₹4,584 ($53.36), yielding annual savings of ₹18,97,484 ($22,086.24), approximately six times the initial investment. These findings demonstrate the potential of MZOCC to reduce operational costs while enhancing occupant comfort.

References:

J.C. Jacob, D. Pandit, J. Sen (2023), Energy-saving potential in Indian open-plan offices using Micro-Zonal Occupant Centric Control (MZOCC), Energy Build. 282 (2023) 112799. https://doi.org/10.1016/J.ENBUILD.2023.112799.
J.C. Jacob, D. Pandit, J. Sen (2024), Investigating enhanced thermal comfort and energy efficiency through strategized airflow in Micro-Zonal Occupant-Centric Control (MZOCC), Energy and Buildings, Volume 318, 2024, 114497, ISSN 0378-7788, https://doi.org/10.1016/j.enbuild.2024.114497.
J.C. Jacob, D. Pandit, J. Sen, (2024). Protocols for planning micro-zones to facilitate occupant-centric control (OCC) to reduce HVAC energy consumption in Indian open-plan offices. Energy Efficiency 17, https://doi.org/10.1007/s12053-024-10271-4.

Keyword: *Energy efficiency, Personalized cooling, CFD*

Exposure to elevated indoor carbon dioxide may affect students cognitive performance: A meta-analytic review of randomized controlled trials

Guangtao Fan * , **Taibo He, Zhongming Huang, Qihong Deng**
Zhengzhou University

Growing evidence suggests that exposure to common indoor carbon dioxide levels (below 5000 ppm) can affect students' cognitive functions. However, the conclusions drawn from various studies are inconsistent. To reconcile these discrepancies, we conducted a systematic review and meta-analysis. This study involved a comprehensive literature search using relevant keywords across several databases, including Web of Science, PubMed, CNKI, and Wanfang Data, with a search deadline set for December 2023. Finally, a total of 11 eligible articles are included in the meta-analysis to assess the impact of short-term exposure to carbon dioxide on students' cognitive performance- specifically attention, executive function and memory. Subgroup analyses considered variations in student age and the duration of exposure. The results indicate that the impact of short-term exposure to indoor CO_2 on students' cognitive performance varies depending on the specific type of cognitive function assessed. When exposed to additional CO_2 concentration ranging from 3000 to 5000 ppm, students' attention (SMD: -0.54; 95% CI: -0.82, -0.25) and executive function (SMD: -0.25; 95% CI-0.42, -0.07)) are more significantly affected in comparison to memory (SMD: -0.19; 95% CI: -0.50, 0.12). In comparison to college students, middle school students exhibit greater susceptibility to the impact of CO_2 exposure on their attention and executive function. Prolonged exposure to elevated concentrations of carbon dioxide may intensify the negative impact on student executive function.

Keyword: *Indoor carbon dioxide, Meta-analysis, Cognitive Performance*

Air Quality Monitoring and Modelling in School Classrooms

Jierui Dong * [1], **Nigel Goodman**[2], **Priyadarsini Rajagopalan**[1]
[1] *RMIT University, Melbourne, VIC 3000, Australia*
[2] *University of Canberra, Australian Capital Territory, 2617, Australia; RMIT University, Melbourne, VIC 3000, Australia*

Background: Air quality in school classrooms can significantly impact children's health and academic performance. Existing evidence shows that classrooms often suffer from poor ventilation, as indicated by carbon dioxide (CO_2) concentrations that exceeding recommended levels. Air quality monitoring and modelling can improve the understanding and control of indoor environment. However, the lack of indoor air quality (IAQ) monitoring data limits the applicability of modelling and management.

Methods: The overarching aim of this study is to develop machine learning-based predictive models for classrooms IAQ. Using low-cost air quality sensors (LCAQS), we will collect indoor and outdoor air pollutant concentrations over four seasons, while also monitoring classroom activities and building operation (e.g., use of windows). Then, we will analyse the pollutants patterns over four seasons. We are currently collecting data in one primary school (PS) and one secondary school (SS).

Results: Preliminary findings reveal that indoor CO_2 concentrations exceeded Australian National Construction Code (NCC) guideline of 850 ppm during class hours, with 16% of the readings surpassing the limit at SS and 25% at PS. Indoor $PM_{2.5}$ at SS was generally lower than outdoors and moderately correlated (Pearson Correlation Coefficient, PCC = 0.64), while at PS, indoor $PM_{2.5}$ was often higher than outdoors and strongly correlated with CO_2 levels (PCC = 0.84).

Conclusion: School air quality require greater attention, especially in PS. This research will continue to collect air quality data across different seasons and locations, then develop predictive models based on long-term monitoring data.

Keyword: *Indoor air quality, Classrooms, Monitoring*

Evaluating the Impact of Building Insulation Materials on Indoor Air Quality, Cognitive Health, and Embodied Carbon in Buildings

Neelima Geetha Archana Madasu * , **Sanand Maddipati, Dongwoo Jason Yeom, Philip Horton, Elham Fini, Ajay Therala**
Arizona State University

Toxic compounds in the 11,000 liters of air an average person inhales daily are often invisible or odorless, making detection difficult. Indoor air pollutant concentrations, where people spend about 90% of their time, are typically 2 to 5 times higher than outdoor levels. Prolonged exposure is linked to cognitive decline, including an increased gap between brain age and chronological age and reduced academic performance. To address decarbonization goals, buildings are increasingly constructed with greater air tightness, using various insulation materials. While this reduces a building's carbon footprint, it can also lead to poor indoor air quality. Off-gassing and airborne particles from occupant activities can degrade air quality. Furthermore, insulation materials like polyurethane, fiberglass, and some bio-based options can emit volatile organic compounds (VOCs) that are harmful to health. Different insulation materials release harmful emissions, which can pose a latent risk to occupant health. Many petroleum-based insulations release VOCs similar to bituminous products, which have negative neurological effects. Bio-based insulation, derived from renewable sources offers a safer alternative with a smaller carbon footprint and fewer harmful emissions, improving environmental sustainability and indoor air quality. This study examines the impacts of insulation materials, including bio-based, polyurethane, and fiberglass, focusing on emissions and cognitive performance. Controlled experiments under varying temperature and humidity conditions assessed thermal resistance, sound absorption, moisture permeability, and air quality. Cognitive performance was tested using the Joggle test, measuring reaction time and memory recall. This analysis provides insight into how insulation materials affect indoor air quality and cognitive health.

References:

Masoumeh Mousavi, Jahangir Emrani, John C. Teleha, Guangming Jiang, Bruce D. Johnson, Ali Shamshiripour, Elham H. Fini, 2024, Journal of Hazardous Materials, Volume 480, https://doi.org/10.1016/j.jhazmat.2024.136048.

Saghar Parikhah Zarmehr, Mohammadjavad Kazemi, Neelima Geetha Archana Madasu, Anthony J. Lamanna, Elham H. Fini, 2025, Resources, Conservation and Recycling, Volume 212, https://doi.org/10.1016/j.resconrec.2024.107906

Keyword: *Indoor air quality (IAQ), volatile organic compound (VOC), cognitive performance*

Residential VOC from Building Materials: Exposures, Health Risks, and Ambient Hazards

Yi Fan * [1], **Jianyin Xiong**[2], **Shaodan Huang**[1]
[1] *Department of Occupational and Environmental Health Sciences, School of Public Health, Peking University, Beijing 100191, China*
[2] *School of Mechanical Engineering, Beijing Institute of Technology, Beijing 100081, China*

Indoor volatile organic compounds (VOCs) from building materials are a significant concern in China, yet their national indoor exposures, health risks, or impact on the ambient remain underexplored. Here, we integrate emission mechanism with tests data on 53 VOC species measured in 279 residences across nine cities, establishing a theoretical-experimental framework linking indoor VOC concentrations to key factors: temperature, humidity, air exchange rate, loading factors, recent renovations, and regional variability. Using Monte Carlo simulations, we estimate the average residential concentrations and health risks of the ten most prevalent VOCs across 31 provincial regions. Residential VOC levels show distinct spatial-temporal variations: concentrations are generally higher in southern than northern regions, with most peaking in summer, while 1,2-dichloropropane, benzene, styrene, and 3-chlorotoluene peak in winter. Cancer and non-cancer risk assessments based on incremental lifetime cancer risks (ILCRs) and hazard quotients (HQs) indicate substantial health burdens, particularly in densely populated urban centers where exposure levels are highest. Beyond indoor environments, VOC emissions from residential materials significantly impact ambient air quality, contributing more in urban than rural areas. Formaldehyde, 1,2-dichloropropane, and cumene account for 13.169%, 14.779%, and 2.475% of total anthropogenic VOC emissions, ranking 3rd, 3rd, and 11th among 99 sources. The estimated O3 and secondary organic aerosol (SOA) formation reaches 110.562 Gg/year and 0.216 Gg/year, respectively, exceeding emissions from nearly half of anthropogenic sources. The pronounced contributions to O3 and SOA formation further underscore the environmental implications of indoor VOC emissions. Our findings highlight the urgent need for targeted mitigation strategies to curb VOC emissions from residential environments, safeguarding indoor air quality, minimizing health risks, and reducing broader atmospheric impacts. Strengthening emission regulations and promoting low-emission building materials are critical steps toward reducing VOC exposure and improving overall air quality.

Keyword: *Indoor air quality; Volatile organic compound; Health risk; Ambient; Sustainability*

Variations in Winter HVAC Utilization Habits Across China's Climate Zones and Their Effects on Arthritis Symptoms in Elderly People

Yan Zhang *

Department of the Built Environment, College of Design and Engineering, Nation University of Singapore, Singapore 117566, Singapore

Environmental differences across climate zones result in significant disparities in arthritis prevalence. This study aims to identify the pathways through which winter HVAC utilization influences arthritis symptoms across different climatic regions and to assess the protective effects of heating practices on elderly health. A cross-sectional study was conducted in four representative cities in China—Chongqing, Beijing, Guangzhou, and Kunming—encompassing 3,579 elderly participants aged 60 to 74. Indoor environmental conditions, HVAC utilization habits, and health status were assessed through structured questionnaires. We employed partial least squares-structural equation modeling to establish the relationships between HVAC utilization habits, indoor environmental exposures, and health outcomes in different cities. Results indicate that the protective effect of HVAC utilization habits on health is strongest in Beijing, while the risk of arthritis due to indoor environmental exposure is greatest in Chongqing. The indirect mediation effect of HVAC utilization via indoor cold exposure (72.32%) was most significant in Chongqing, while indoor damp exposure (44.71%) had the strongest effect in Guangzhou, and indoor air pollution (22.19%) in Beijing. The overall strength of indirect effects follows the order: Chongqing > Guangzhou > Beijing > Kunming. This study demonstrates that region-specific HVAC strategies tailored to climate characteristics are beneficial for alleviating arthritis symptoms in elderly patients, especially in high risk zones.

Keyword: *Multi-city, HVAC utilization habits, Indoor environmental exposure*

Research agendas for technologies to suppress indoor pathogen transmission

Aman Patel * , **James Montavon, Richard Williamson, David Carel, Jacob Swett**
Blueprint Biosecurity

Reducing the concentration of pathogens in occupied indoor spaces is essential for preventing and responding to future pandemics, but traditional ventilation and filtration alone may not provide enough air cleaning capacity. Via technology landscape mapping, we identified two promising technologies that could be used to supplement indoor effective air change rates: far-UVC light and triethylene glycol vapor. Current evidence suggests that far-UVC light can be directed to a whole room and will inactivate pathogens by causing nucleic acid and protein damage in proportion to fluence rates. Triethylene glycol vapor dissolves into ambient moisture-containing droplets and (once in the liquid phase) dehydrates pathogens, increases membrane fluidity, and disrupts protein structure.

However, both of these technologies require additional research on safety, efficacy, scalability, and real-world implementation before they will be used for pandemic prevention or response. Via interviews and conversations with stakeholder groups, we have identified minimum research agendas needed to further investigate each technology. For far-UVC, our proposed research agenda includes studies on pathogen susceptibility in human respiratory aerosols; standardization of lab-based efficacy methodologies; investigation of novel effects distinct from conventional UV; assessment of indoor air quality impacts and mitigation strategies; and modeling to guide practical deployment. For triethylene glycol vapor, our proposed research agenda includes studies on efficacy under different environmental conditions; degradation product hazards in buildings; accuracy of vapor concentration measurement methods; effects on filtration media; pathogen reduction in high-risk settings; toxicology; and rapid industrial production. Our work now focuses on aligning the scientific community around these research agendas.

Keyword: *Far-UVC, Triethylene glycol, Air disinfection*

A Blueprint for Far-UVC to Reduce Airborne Pathogens in the Built Environment

James Montavon * [1], **Richard Williamson**[1], **Bryan Lehrer**[1], **Harry Koos**[2], **Jacob Swett**[1]
[1] *Blueprint Biosecurity, Washington, DC, USA*
[2] *Stanford University School of Medicine, Stanford, CA, USA*

Protecting people from infectious aerosols in their indoor air is a remarkable opportunity for improving public health, but comes with significant challenges. Recent standards, such as ASHRAE 241, set equivalent clean air delivery rate targets to reduce airborne transmission risk. While increased ventilation and added filtration can meet these targets in some settings, many environments—such as classrooms, gyms, and restaurants—require impractically high ventilation rates to achieve sufficient infection risk mitigation. This gap highlights the need for additional strategies to supplement traditional air quality interventions. Far-UVC – an emerging type of germicidal UV light – could be one of these strategies.

To develop an actionable roadmap for far-UVC disinfection, we conducted extensive literature reviews, semi-structured interviews with over 100 experts from academia, industry, government, healthcare, and the non-profit sector, and performed parametric modeling using the Wells-Riley infection risk framework. This multidisciplinary effort evaluated far-UVC's potential to inactivate pathogens in diverse settings, focusing on technical domains such as disinfection efficacy, photobiological safety, materials compatibility, and indoor air chemistry.

When integrated with other air cleaning technologies like ventilation and filtration, far-UVC offers a cost-effective and energy-efficient way to achieve equivalent clean air delivery rate targets, particularly in spaces where increased ventilation is impractical. Our research emphasizes how far-UVC offers a cost-effective and energy-efficient alternative to achieving equivalent clean air delivery rate targets. Our poster will present findings on how building characteristics like airflow, occupancy, and space constraints influence optimal far-UVC deployment.

Keyword: *Far-UVC, Germicidal UV, Airborne Transmission*

Indoor Air Quality in Indian restaurants and its perception

Pooja Agarwal * , **Jay Dhariwal**
Indian Institute of Technology Delhi

People typically spend 1-2 hours dining in a restaurant, and current trends indicate that 80% of Indians eat out at least 5-6 times a month. Therefore, it is crucial to understand the potential links between occupants perception of IAQ and physical exposure to indoor pollution in restaurants. In the current study, we evaluated the perception of customers and staff in two air-conditioned restaurants in Delhi in the winter season through point-in-time surveys and indoor objective measurements of climate parameters (DBT and RH) and air pollutants (CO_2 and $PM_{2.5}$). None of the restaurants used air purification techniques. We evaluated the occupants' perception on a 7-point scale and conducted sensor-based monitoring to measure air pollutants. About 88% of the occupants (inspectors, staff, and customers) expressed dissatisfaction with the indoor air quality, despite finding the thermal conditions to be acceptable even when the restaurants were operating at less than 50% capacity. A negative correlation was found between the time spent by customers in the restaurants and pollution load. This could be attributed as a consequence of increased levels of CO_2 and $PM_{2.5}$, as well as decreased levels of acceptability of indoor air quality (IAQ). The occupants did not perceive food smells as a cause of dissatisfaction. The study also observed a statistically significant inverse relationship between the perception of IAQ and the perception of thermal comfort.

References:

1. Kim, J., Jang, M., Choi, K., & Kim, K. (2019). Perception of indoor air quality (IAQ) by workers in underground shopping centers in relation to sick-building syndrome (SBS) and store type: a cross-sectional study in Korea. BMC Public Health, 19(1). https://doi.org/10.1186/s12889-019-6988-6
2. Langer, S., Ramalho, O., Ponner, E. L., Derbez, M., Kirchner, S., & Mandin, C. (2017). Perceived indoor air quality and its relationship to air pollutants in French dwellings. Indoor Air, 27(6), 1168–1176. https://doi.org/10.1111/ina.12393
3. Zhang, X., Wargocki, P., Lian, Z., & Thyregod, C. (2016). Effects of exposure to carbon dioxide and bioeffluents on perceived air quality, self-assessed acute health symptoms, and cognitive performance. Indoor Air, 27(1), 47–64. https://doi.org/10.1111/ina.12284

Keyword: *indoor air quality, occupants perception, surveys*

THE EFFECT OF ACOUSTIC BARRIERS ON DAYLIGHT IN A CLASSROOM

Asvika K S * , **Subbaiyan G**
Department of Architecture, National Institute of Technology, Tiruchirappalli

Traffic noise is a significant sound pollutant in urban environments. Previous studies have identified that schools near the roadside are severely impacted by Traffic noise, affecting the teaching-learning process. The windows required for natural lighting and ventilation are the weakest part of the façade for sound insulation. So, providing acoustic shades in the window becomes a solution, but these shades affect the daylight that enters the classroom. This leads to questions like how a shading system can improve noise reduction and daylight in the classroom, how the tilt of the fins affects the Noise reduction and daylight in the classroom, and how we can balance the daylight and acoustics by retrofitting the fins and WWR.

An educational institution in an urban area adjacent to a busy road is considered for the study. Simulations were carried out for the indoor noise level of an empty classroom, with varying horizontal louver angles and WWR for different floors, with the sound pressure level of outdoor noise fixed at 70dB (NBC 2016). A layer of sound-absorptive material was provided on the bottom side of the louver, reducing the noise ingress that made most cases comply with IS 8827 for the background noise level threshold for educational institutions.

The upward tilt of louvers had lower SPL (A) than the case without shade. The daylight of the room drastically reduces as the angle of the louver goes higher. It was noticed that an angle of -15° to 60° complies with the LEED standard of daylighting.

Keyword: *Daylighting, Acoustic Louvers, Classrooms*

Multisensory synergies and conflicts in occupant adaptation to semi-outdoor environments

Kun Lyu * [1], **Arianna Brambilla**[2], **Anastasia Globa**[2], **Richard de Dear**[2]
[1] *École Polytechnique Fédérale de Lausanne*
[2] *University of Sydney*

There is a growing emphasis on implementing urban heat mitigation and thermal adaptive design strategies to enhance quality of (semi-)outdoor spaces and promote human well-being. In addition to the effectiveness in improving thermal comfort, these measures may bring corresponding changes of other sensory environments for the inhabitants, such as soundscapes and visual aesthetics. Conflicts between sensory modalities can arise when occupants' behavioural adaptation to thermal conditions, such as relocating, compromises other sensory experiences. This issue is rarely considered in the research and practice to date. This study aims to understand the dynamic processes of occupant behavioural adaptation to semi-outdoor and outdoor environments considering synergies and conflicts between multisensory environmental qualities. An experiment with human subjects from Australian and Chinese cultural backgrounds was conducted in a Virtual Reality integrated climate chamber, where the visual, auditory, and thermal conditions of two semi-outdoor environmental scenarios (UTCI= 31.4°C and 28.4°C) were simulated. During the environmental exposure, participants were able to freely explore the environment and choose their favourite locations and orientations. Their subjective experiences were surveyed with questionnaires and interviews; behaviours inside VR were monitored. Key multisensory environmental properties underpinning occupants' preferences emerged from the analysis, including prospect and refuge, and shade and sunshine. The study revealed how these multidomain factors interact to shape occupants' adaptive behaviour and environmental satisfaction. Findings provides a nuanced understanding of the multisensory adaptive processes in operation in semi-outdoor environments. A holistic approach to multisensory interactions is essential for designing semi-outdoor spaces that enhance urban liveability and promote well-being.

Keyword: *thermal pleasure, adaptive behaviour, multidomain conditions*

Thermal radiation in high-temperature environments exacerbates the damage to human thermal responses and cognitive functions

Yingying Zhao * [1], **Weiwei Liu**[1], **Wargocki Pawel**[2]
[1] *School of Architecture and Art, Central South University, Changsha, China.*
[2] *echnical University of Denmark, Kongens Lyngby, Denmark*

Workplaces with coexisting high temperature and thermal radiation impose significant thermal stress, severely impairing workers' physiological, psychological, and cognitive functions. However, existing research remains insufficient, and current safety standards lack defined safety thresholds for thermal radiation intensity. This study conducted climate chamber experiments to systematically evaluate 36 participants under three conditions: 26°C (baseline), 32°C (high temperature alone), and 32°C with 2 kW/m^2 thermal radiation (combined exposure) during 135-minute sessions. Results demonstrated that, compared to 26°C, both 32°C and combined exposure significantly increased core temperature, skin temperature, heart rate, low-to-high frequency ratio (LF/HF), weight loss, and thermal sensation votes, while reducing peripheral oxygen saturation (SpO_2), thermal comfort and thermal acceptability ($P<0.05$). Acute health symptoms related to cognition markedly worsened, including impaired concentration, mental fatigue, and heightened negative emotions ($P<0.05$). Cognitive assessments revealed performance declines in the Stroop test (semantic interference), visual perception tasks, Tsai-Partington test (arousal level), and working memory tests (1-back and high-low load tasks). These findings provide critical theoretical foundations for establishing updated safety standards in high-temperature and thermal radiation environments.

Keyword: *physiological parameters, psychological parameters, cognitive performance*

The effect of variations in self-pollutant concentration induced by changes in ventilation volume on the cognitive performance of individuals wearing masks.

Xinpeng Yang * [1], **Weiwei Liu**[1], **Pawel Wargocki**[2], **Xiaoyue Lang**[1]
[1] *School of Architecture and Art, Central South University*
[2] *International Centre for Indoor Environment and Energy, Department of Environmental and Resource Engineering, Technical University of Denmark*

Wearing masks reduces the risk of contracting respiratory infectious diseases, but it also introduces additional impacts. However, few studies have systematically investigated whether current indoor temperature standards remain applicable in the context of mask-wearing, particularly when inadequate ventilation leads to elevated self-pollutant concentrations. To address this issue, we conducted climate chamber experiments. The experimental design included two self-pollutant concentration levels (1500 ppm and 3000 ppm), two temperature conditions (26 °C and 28 °C), and two mask-wearing conditions (mask-wearing and non-mask-wearing). Relative humidity was maintained at 60% across all eight experimental conditions. During a 190-minute exposure period, the cognitive performance, physiological responses, and subjective feelings of 16 male and 16 female participants were recorded while they remained seated. The results revealed a coupling effect between self-pollutant concentration and mask-wearing on cognitive performance. Specifically, when participants wore masks, their accuracy in multiplication and addition tasks significantly declined as the temperature increased from 26 °C to 28 °C. In contrast, when participants did not wear masks, the accuracy of high-load tasks significantly decreased as the self-pollutant concentration increased from 1500 ppm to 3000 ppm at 28 °C, and acute health symptoms, such as dry throat, also worsened with the increase in pollutant concentration. These findings suggest that reducing ambient temperature may help mitigate the negative impact of mask-wearing on cognitive performance in situations of poor ventilation.

Keyword: *Protective masks, Cognitive performance, Self-pollutant*

Assessment of Energy and Daylight Models Across Building Typologies for High-Performance Design

<u>Deepashree KS</u> * [1], **Nikita Khatri**[2]
[1] *CEPT University, Ahmedabad, India*
[2] *Ecofirst Services Limited, Bengaluru, India*

The intersection of energy performance and natural daylighting in buildings is crucial for achieving high-performance structures. Energy and daylight modelling are essential tools for analysing how building functions influence energy use and daylight availability, enabling the optimization of energy and occupant comfort.

This research aims to analyse the impact of building functions on energy demand and daylight distribution, examining how occupancy patterns, spatial layouts, and design choices affect performance. By leveraging simulation tools, the study intends to provide practical design recommendations for optimizing window placement, shading strategies, interior layouts, and HVAC systems to reduce energy consumption and enhance thermal comfort in buildings.

As cities and buildings grow increasingly complex, with diverse functional requirements, the need for such assessments is becoming more critical. Exploring energy and daylight modelling through the lens of building function offers valuable insights into how different spaces interact with natural and artificial energy systems.

Furthermore, establishing a correlation between energy use intensity (EUI) ranges of various building typologies with their respective climate types can help formulate standard practices to ensure measurable improvements in overall building performance.

By addressing these challenges, this research contributes to the development of smarter building designs that are not only energy-efficient but also promote healthy, comfortable living environments for the future.

Keyword: *High-performance buildings, energy & daylight simulations, Building typologies*

Evaluation of Environmental Interventions on Mucus Clearance Efficiency in Cystic Fibrosis Patients Using CFD Simulations

Yijia Chen * [1], **Weiqi Guo**[1], **Mengjie Duan**[2], **Li Liu**[1]
[1] *Department of Building Science, Tsinghua University, Beijing, 100084, China*
[2] *Vanke School of Public Health, Tsinghua University, Beijing, 100084, China*

Cystic fibrosis (CF) patients frequently experience bronchial mucus obstruction, which is exacerbated by indoor environmental pollutants, leading to impaired mucus clearance and acute exacerbations. This study integrates clinical samples from CF patients with computational fluid dynamics (CFD) to evaluate the efficacy of expectorant therapies, such as chest percussion and aerosolized medications, in enhancing mucus clearance during coughing.

Mucus samples were collected via bronchoscopy, and their surface tension and viscoelastic properties were analyzed using a rheometer under various environmental conditions. Cough-flow dynamics were measured with a spirometer and used as boundary conditions in CFD simulations, with patient-specific lung geometries derived from CT scans. A personalized numerical framework was developed to assess mucus clearance strategies, incorporating individual physiological and environmental factors.

The Eulerian Wall Film-Discrete Phase Model (EWF-DPM) was utilized to explore the interplay between mucus properties, airway morphology, and exhalation dynamics. Simulations highlighted distinct differences in mucus physical properties and clearance dynamics between CF patients and healthy individuals, particularly in bronchiectatic regions. The study quantified mucus dislodgment and its correlation with bronchial obstruction, providing insights into clearance efficiency under different environmental exposures.

These findings facilitate the evaluation of environmental interventions on expectoration efficiency, guiding personalized treatment strategies. This research advances non-invasive interventions for chronic respiratory disease management, offering a foundation for optimizing therapeutic approaches and ultimately improving the quality of life for CF patients.

Keyword: *Cystic fibrosis, Computational fluid dynamics (CFD), Expectorant efficiency*

Development of a Non-Invasive Activity Classification Model for Bathrooms Using Environmental Data

Jun-Sub Kim * , **Jeong Won Kim, Hyeun Jun Moon**
Dankook University, Yongin, Republic of Korea

This study focuses on the development of a non-invasive activity classification model for bathrooms using environmental data. Bathrooms, while essential in daily life, pose significant risks for vulnerable populations - including the elderly and individuals with disabilities - due to the potential for falls and other emergencies. By utilizing environmental data - such as temperature, humidity, CO_2, PIR, illuminance, VOCs, and particulate matter - this study avoids the privacy concerns of cameras and wearable devices while effectively monitoring user activities.

Environmental data were collected from apartments in Seoul, South Korea, from February 2023 to May 2024. Seven predefined activities - handwashing, showering, cleaning, toileting, laundry loading, bathing, and laundering - were analyzed using labeled activity data to uncover patterns and correlations. A Random Forest-based classification model was developed to categorize these activities. The model incorporated a 10-minute window before and after each labeled action, capturing the temporal context of activities. Preprocessing was applied to handle missing values and improve data reliability. The model achieved over 90% accuracy in classifying activities, with CO_2 and temperature identified as the most influential features through feature importance analysis.

This research demonstrates the practical application of non-invasive environmental data for accurate activity classification in bathrooms. Beyond high accuracy, the model offers a foundation for detecting abnormal patterns - such as extended inactivity - which may signal emergencies. By integrating this classification model into Ambient Assisted Living (AAL) systems, the framework provides a scalable and privacy-conscious solution for creating safer and smarter environments for vulnerable users.

References:

Haagsman, E. M., de Bruin MSc, R., & Groothuizen, T. J. Falls in the toilet environment: a study on influential factors.
Yoon, Y. R., Lee, Y. R., Kim, S. H., Kim, J. W., & Moon, H. J. (2022). A non-intrusive data-driven model for detailed occupants' activities classification in residential buildings using environmental and energy usage data. Energy and Buildings, 256, 111699.

Keyword: *Activity classification, Environmental data, Ambient Assisted Living (AAL) systems*

Paradigm Shifts in Indoor Air Quality: Insights from the 2025 Stanford Forum on Sustainable and Healthy Buildings

Milana Trounce * [1], **Lidia Morawska**[2]
[1] *Stanford University School of Medicine*
[2] *Queensland University of Technology (QUT)*

In March 2025, Stanford University hosted the Forum on Sustainable and Healthy Buildings, a high-level interdisciplinary event aimed at exploring how Indoor Air Quality (IAQ) performance standards can be feasibly implemented in ways that also support energy efficiency and climate resilience. Co-chaired by Distinguished Professor Lidia Morawska (QUT/WHO) and Dr. Milana Trounce (Stanford), the forum convened 20+ experts from academia, government, building design engineering, real estate, and broader industry. It was organized in response to the growing recognition—supported by a recent Science consensus paper (Morawska et al, 2024)*—that clean indoor air is essential infrastructure in the face of climate-driven wildfires, airborne infectious disease threats, and broader public health challenges.

Forum participants engaged in structured roundtable discussions on the feasibility of IAQ performance standards, including technical pathways for compliance in new and existing buildings, costs of implementation, interaction with energy and emissions targets, and strategies for market transformation. A particular emphasis was placed on real-world barriers to adoption—including measurement, retrofitting, and design limitations—and how these can be overcome through regulation, certification, technical innovation, pilot programs, and market demand.

Key outcomes included the formation of multi-stakeholder working groups focused on: (1) model IAQ performance standards, (2) pilot implementation projects, (3) integrated policy and market frameworks, and (4) cross-sector education, dissemination, and research.

This abstract will present forum insights and early outputs from these working groups, with the goal of advancing international dialogue and collaboration on practical, scalable solutions for clean, healthy, and low-carbon

Keyword: *IAQ performance standards, building resilience, air quality policy*

The Future of IEQ Regulations - A View from North America

William Bahnfleth *
The Pennsylvania State University

While the effects of indoor environmental quality (IEQ) have been studied for a very long time, the last half-century has been a particularly productive period during which great advances in understanding of how air quality, thermal environment, light, sound/vibration affect human health and well-being. Science has advanced from focusing predominantly on perceived quality of the indoor environment and relationship between exposures and harm parameter by parameter to consideration of quantifiable benefits of improved IEQ and, more ambitiously, attempts to define metrics that holistically describe the cumulative effects of all indoor environmental exposures. To an extent, standards have adopted the findings of this rapidly expanding body of research, but to a surprisingly limited degree. Furthermore, regulations, i.e., mandatory requirements are very few compared with what evidence suggests would be justified. This paper will provide an assessment of the current status in North America with some discussion of historical context. The most urgent gaps in knowledge, technology, standards, and regulations and barriers to progress will be summarized. A deeper discussion of indoor air quality will illustrate the key issues related to a specific IEQ parameter. The paper will conclude with speculation based in the author's experience, of how achieveable goals and a path to reach them can be defined.

Keyword: *Standards, Regulations, IEQ*

Thermal Comfort in Buildings of Nepal: Policies, Practices, and Potential

Sangita Thapa * , **Abhishek Karn**
Institute of Engineering ,Pulchowk Campus, Lalitpur, Nepal

With the topological and climatic diversity of Nepal, ensuring thermal comfort has always been critical factor in building design and is traditionally reflected in the vernacular architecture of the region. However, with the frequent earthquakes and changing construction practices, these climate responsive design are gradually being replaced by the modern construction practices such as concrete and glass structures. Regardless of the temperature reaching upto 45^0 celcius in Terai region or -25^0 celcius in the Himalayan region, buildings are being designed without consideration of local climate, adversely affecting the indoor thermal comfort and increased energy consumption to compensate for this discomfort.

To address these issues, several initiatives have been undertaken which include Green Homes by UN Habitat in Nepal, ongoing Green Building Guidelines by the World Bank, Energy Efficient building policy by BUILDING Energy Efficiency in Nepal (BEEN project). The BEEN project has successfully incorporated energy efficiency provisions into the building permit system of 24 municipalities. Despite these efforts, thermal comfort is inadequately addressed in National Building Code of Nepal. Because of this, many newly constructed buildings continue to exhibit poor indoor thermal performance, leading to discomfort for occupants and increased energy demand.

Hence, this paper provides us valuable insights into the current energy efficiency policies for buildings in Nepal and highlights key gaps that need to be addressed to design more thermally comfortable and healthy buildings. By referencing international thermal comfort standards, energy-efficient building codes, and global energy rating systems, it outlines a strategic path forward for implementation in Nepal.

Keyword: *Thermal comfort, energy efficiency, passive design, policy implementation, policy advocacy*

Critical analysis of net zero pathways for affordable housing in India

Surjyatapa Ray Choudhury * , **Nithya Ramesh**
Jana Urban Space Foundation

India is at a tipping point of climate vulnerability. Indian cities are particularly at high risk, given that 40% of India's projecte infrastructure for 2040 is yet to be built. A significant portion of this demand lies in the residential sector. The urban residential sector was adding 0.52 billion sqm floor space annually in 2020. By 2050, this is expected to go up by 82%, adding 0.95 billion sqm annually. 96% of this demand comes from the low income group (LIG) segment. There is thus an urgent need to ensure that India's net zero transition is affordable for the LIG segment.

We look at five pathways for emission reduction – (i) renewable energy transition, (ii) energy efficiency, (iii) sustainable materials, (iv) sustainable construction, (v) building retrofit. We examine regulatory instruments, policies and schemes, RD&D framework through the Availability-Accessibility-Quality framework, examining if (i) solutions are available, (ii) solutions are accessible/ affordable, and (iii) quality of solution is high w.r.t. emission reduction. Findings reveal that some solutions are affordable, but consumers are disincentivized in the presence of contradictory sub-national schemes. Additionally, where affordable housing has been incorporated in the solutions, there is a dilution in impact in terms of GHG emissions. Based on the findings, we propose three key thrust areas – defining what is green, converging policies with building regulations, and sustainable procurement, to ensure an affordable net zero future for India.

References:

BEE, & AEEE. (2022). State Energy Efficiency Index 2021-22.
Bureau of Energy Efficiency. (n.d.). Retrieved September 1, 2023, from https://saathee.beeindia.gov.in/Home/EEBC_Residential#:~:text=ENS%20cells%20shall%20be%20established,the%20voluntary%20stage%20of%20adoption
MoEFCC. (2024). Fourth Biennial Update Report to The United Nations Framework Convention on Climate Change.

Keyword: *Affordable housing, Policy implementation, Net Zero Built Environment*

Cleaning, Restoration, and Inspection Standards: What Are They, and Why Do We Need Them?

Mili Washington *

IICRC

This abstract touches on the different aspects of standards development - what they are, how they should be developed, and why they are important for Healthy Buildings and the occupants within. Presentation to include the following:

Essential Requirements for Standards Development: As an American National Standards Institute (ANSI) Accredited Standards Developer (ASD), IICRC adheres to the following principles for due process: Openness, Lack of Dominance, Balance, Consensus, Coordination and Harmonization, Notification of standards development, Consideration of views and objections, and Appeals.

Standard of Care: defined as the practices that are common to reasonably prudent members of the trade who are recognized in the industry as qualified and competent.

Developing New and Revised Standards: summarized overview of the process of developing and revising Standards.

Why is it important to follow and participate in standards? Simply put, standards work to create healthy buildings, and in turn, healthy occupants. Standards are created to find the best possible method of performing a specified task - as presented by industry experts, debated, agreed upon through consensus, and finally reviewed extensively by industry peers to ensure they are fair and work as intended.

Standards to be introduced include: performance Standards for mold remediation; water damage restoration; fire and smoke damage restoration; cleaning, maintanence, and restoration of the built environment; and cleaning of the built environment for infection control. These Standards can be used globally via international adoptions, and go a long way to ensure a healthy built environment; presentation is not intended to cover failure of these Standards.

Keyword: *Standards, Standards Development, American National Standards Institute*

Towards more holistic rating systems for green and healthy buildings

Vaibhav Lawange * , **Sameer Patel**
Indian Institute of Technology Gandhinagar, India

Currently, 250 billion square meters of built-up area exist globally, demonstrating the extent of urban development and the increasing demand for infrastructure. This trend is particularly evident in developing countries such as India, which invested 5.6% of its GDP in 2020, ~USD 200 billion, for infrastructural development. The building sector currently consumes ~34% of global energy and ~40% of natural resources. Consequently, it accounts for more than 10 $GtCO_2$, nearly ~37% of global CO_2 emissions. Simultaneously, buildings govern occupants' health and well-being since people spend ~80% of their time indoors. Therefore, environmental factors affect human health directly and indirectly throughout the building's life cycle. The World Health Organization reported 24% of global deaths due to environmental factors in 2020, with 3.2 million fatalities owing to poor indoor air quality.

Considering the building stock is poised to keep growing in line with urbanization, it is critical to focus on efficient and healthy buildings to minimize environmental and human health impacts. To promote sustainable building practices, Green Building Rating Systems (GBRS) were initiated by Green Building Councils. While popular, the underlying structure of GBRSs scoring and weights assigned to sustainability criteria appears not to correlate well with the environmental impacts of those criteria. This work attempts to revise the scoring and weightage system of GBRS by reassessing each criterion's appraisal's weightage, determining reduced carbon footprint as a quantification unit.

Keyword: *Green Building Rating System, Life Cycle Assessment, Weightage Distribution, Healthy Building, Sustainable Buildings*

Addressing Indoor Air Pollution in Mogadishu, Somalia: Policy Gaps and Future Solutions

Abdifatah Munye * [1], **Zamzam Osman**[1], **Abdullahi Sheikh Ali**[2], **Abdulshakur Diiso**[3]
[1] *Faculty of Engineering, Department of civil Engineering, Abrar University, Mogadishu, Somalia*
[2] *Department of Civil Engineering, Indian Institute of Technology, Delhi Hauz Khas, New Delhi, 110016*
[3] *Faculty of Geoscience and Environment, Almaas University, Mogadishu, Somalia*

The indoor air pollution is a major issue in Somalia and has a direct linkage with household level energy use. Indoor air pollution due to solid fuel burning is a central issue in Somalia, considering that 58% of all households burn firewood for cooking. It is assumed that every household consumes approximately 3 kg of wood a day for cooking; a total of 1.5 million tons of wood are burned a year across Somalia, with an equivalent of 2.7 million tons of CO_2 emissions. The estimated annual per capita emission of 0.2 tons of CO2-equivalent of biomass fuel burning in Somalia causes substantial adverse health effects among the population. In Mogadishu, Somalia, indoor air pollution (IAP) is a significant yet understudied aspect of environmental and public health. Biomass fuel use, such as firewood and charcoal, considerably deteriorates indoor air quality, leading to respiratory illnesses and other ailments. Despite increasing awareness of the detrimental effects of air pollution, Somalia does not have an overarching policy framework for controlling IAP. This research explores the key sources of indoor air pollution, the health impacts, regulatory gaps, and remedies. Through a review of prevailing legal regimes, global best practices, and empirical research, this research proposes a comprehensive multi-sectoral approach to mitigating indoor air pollution in Mogadishu. This entails public awareness, clean energy promotion, legislative change, and urban planning. Implementation of these measures can make a significant difference in Somalia's indoor air quality and public health.

References:

Abas Mohamed Sharif, Vice Rector for Academic Affairs Contact: abas@abrar.edu.so tell: +252 61 5553205
Eng Ahmed Mohamed Hassan Dean of Faculty of Engineering and Technology At Abrar University Contact: Dean.eng@abrar.edu.so tell: +252 61 5652758

Keyword: *Indoor air pollution, Indoor air quality, Health*

Title: Policy and Regulatory Impacts of Electric Vehicle Adoption on Indoor Air Quality in Developing Economies: The Case of India

Sitha MahaLakshmi P * [1], **Narayankumar Sreekumar**[2]

[1] *PhD Research Scholar, School of Planning and Architecture Vijayawada, Andhra Pradesh, India.*

[2] *Senior Specialist, Industrial Energy Efficiency, Department of Energy, Government of AbuDhabi*

Transitioning from internal combustion engine (ICE) vehicles to electric vehicles (EVs) offers a significant opportunity to improve urban air quality, with ripple effects on indoor air quality in buildings. This study explores the regulatory challenges and opportunities in India's EV transition and their impact on indoor air quality. By analyzing global case studies and regulatory frameworks, key factors adaptable to the Indian context are identified. The research investigates the current state of EV adoption in India, existing policies, and areas for improvement.

The study analyzes how overcoming challenges in infrastructure development, financial incentives, manufacturing, and public awareness can accelerate EV adoption and enhance outdoor and indoor air quality. Findings show that this transition can reduce urban air pollution by up to 30%, with significant health benefits. For example, California's Zero Emission Vehicle Program decreased particulate matter (PM2.5) by 5% and respiratory-related hospital admissions by 3%. China's New Energy Vehicle Policy led to a 20% reduction in NOx emissions and a 15% reduction in PM2.5 levels, improving respiratory health. Norway's EV incentives resulted in a 30% reduction in NOx emissions and a 25% decrease in PM2.5 levels, correlating with a 10% reduction in respiratory diseases.

The study examines the potential implications of reduced vehicle emissions on building ventilation systems and occupant health. The findings leads to insights and recommendations for policymakers, urban planners and building designers in creating sustainable and healthy built-environment in India. This research contributes to the growing discourse on sustainable urban-development and public health in rapidly urbanizing regions.

References:

• Harvard Kennedy School: "Charging the Future: Challenges and Opportunities for Electric Vehicle Adoption" (2018).
• American Journal of Public Health: "The Impact of California's ZEV Program on Air Quality and Public Health" (2020).
• Lancet Planetary Health: "Norway's EV Incentives and the Resulting Public Health Benefits" (2021).

Keyword: *Electric Vehicles (EVs), Indoor Air Quality, Regulatory Challenges*

Effects of Road Traffic Noise and PM2.5 Pollution and Meteorology on Office Employees Mean Arterial and Pulse Pressure

Argha Kamal Guha, Sharad Gokhale

First and corresponding author

Argha Kamal Guha
Former Ph.D Research Scholar
Department of Civil Engineering
Indian Institute of Technology Guwahati
Guwahati; Assam; India
Pin-781039

Present Affiliation
Assistant Professor
Department of Civil Engineering
Adamas University; Kolkata; West Bengal; India
Pin-700126
E-mail: arghakamalguha@gmail.com/ arghakamal.guha1@adamasuniversity.ac.in

Second author

Sharad Gokhale
Professor
Department of Civil Engineering
Indian Institute of Technology Guwahati
Guwahati; Assam; India
Pin-781039
E-mail: sharadbg@iitg.ac.in

Abstract

Urban road traffic noise and air pollution are increasingly recognised as silent threats to cardiovascular health, yet research remains limited in rapidly developing cities. This study investigates the effects of road traffic noise (Traffic Noise Index and Noise Pollution Level), PM2.5 concentrations, and meteorological variables (temperature and relative humidity) on mean arterial (MAP) and pulse pressure (PP) of roadside office employees in Guwahati city. The noise level, PM2.5, and meteorological parameters were monitored (January-December, 2019) inside a roadside public office building. The questionnaire-based health survey (blood pressure measurement) was carried out to assess the office employees' ($N = 108$) health status. The observed data were analysed using various statistical techniques (descriptive statistics and Spearman's rho correlation). The results revealed that both PM2.5 and noise indices exceeded the WHO guidelines. While overall MAP and PP values remained within normal ranges, elevated PP was significantly associated with relative humidity inside the office, increasing age, and perceived roadside annoyance of the office employees. These findings highlight the complex interplay between urban environmental stressors and employees' cardiovascular health. The study underscores the importance of integrating pollution control, indoor climate regulation, and psychological well-being into the design of healthier and more productive urban workplaces.

Keywords Traffic Noise Index; Noise Pollution Level; PM2.5; Annoyance; Mean Arterial Pressure; Pulse Pressure

Assessment Of Relationship between Thermal Comfort, Occupant Behavior, and Classroom Characteristics in Higher Education Institutions A Case of College of Engineering Trivandrum

Diya Pradeep * , **Dr.Deepa Rani R**
College of Engineering Trivandrum,kerala,India

Thermal comfort refers to how well individuals feel conditioned in their environment, influenced by factors like air temperature, humidity, air movement, clothing, and activities. Achieving thermal comfort is challenging due to varying occupant preferences, activities, and architectural constraints. This study explores the relationship between thermal comfort, occupant behavior, and classroom characteristics, including window-to-wall ratio (WWR), room volume, and ventilation strategies, in naturally ventilated higher education institutions.

A structured methodology is employed, integrating field measurements, occupant surveys, and statistical analysis. Environmental data, including air temperature, globe temperature, relative humidity, and air velocity, is recorded using instruments like the Extech HT30 Heat Stress WBGT Meter. Simultaneously, occupant surveys assess thermal perception using ASHRAE's 7-point scale, while behavioral observations document adaptive strategies such as opening windows and adjusting clothing. The collected data is analyzed using statistical correlation and regression models to establish key relationships.

The findings contribute to a conceptual model that serves as a decision-making tool for architects and policymakers, optimizing classroom design for improved thermal comfort. However, the study is limited to a small number of classrooms within two blocks, which may not fully represent diverse classroom designs. Additionally, data collection was conducted within a specific timeframe, potentially overlooking seasonal variations. Future research should expand the sample size, incorporate more institutions, and conduct year-round measurements to develop a more comprehensive thermal comfort model suited for tropical climates, ensuring sustainable and energy-efficient classroom designs.

References:

• ASHRAE Hand book. 2020.
• Dr Priyam Tewari, Associate Fellow, TERI Dr Rana Veer Pratap Singh, Research Associate, TERI. "Thermal comfort prescription for cooling dominated Indian residential buildings." 2021.
• Guenther, Sebastian. Simscale. 2024.

Keyword: *Thermal Comfort, Occupant Behavior, Classroom characteristics*

Behavioral Acumen for Sustainable Living: Examining Residential Energy Use in India

Deepali Ranavainshaw * , **Shankha Pratim Bhattacharya**
Department of Architecture and Regional Planning, IIT Kharagpur, Kharagpur, India

Understanding occupant behaviour is crucial in addressing the energy performance gap in residential buildings as actual energy consumption often deviates from predicted values due to variations in occupant behavior. This study aims to analyze the occupants' energy consumption habits, appliance usage patterns, and responsiveness to behavioral interventions in residential buildings.

Research questions:

1. What factors influence the occupants' behaviors to decide the magnitude of energy consumption?

2. Do behaviour nudges play a role in altering energy consumption habits, and to what extent do they impact energy savings?

Methods:

The study conducted a structured questionnaire survey in residential buildings, focusing on behavioural influences on energy use and addressing key aspects such as behaviour nudging, rebound and prebound effects, and awareness of energy-conscious habits among occupants. It assessed factors such as awareness of energy-efficient practices, energy consciousness, and the impact of economic and social factors on consumption behaviors.

Results:

Energy efficiency is mostly dependent on user participation and accessibility, according to the study. Predicted and actual performance differ. Family size, gender, age, education, and other socioeconomic characteristics affect household energy usage. High environmental awareness and government incentives reduced energy use. However, rebound effects occurred when occupants used more energy-intensive equipment to offset efficiency gains.

Conclusions:

The study stresses the importance of combining behavioral and technical solutions for sustainable energy consumption. These findings can help policymakers, architects, and energy planners understand energy consumption behavior and devise tailored methods to address behavioral disparities.

References:

Andersen, R. K., Toftum, J., Andersen, K. K., & Olesen, B. W. (2009). Survey of occupant behavior and control of indoor environment in Danish dwellings. Energy and Buildings, 41(1), 11-16.
Santin, O. G., Itard, L., & Visscher, H. (2009). The effect of occupancy and building characteristics on energy use for space and water heating in Dutch residential stock. Energy and Buildings, 41(11), 1223-1232.
Frederiks, E. R., Stenner, K., & Hobman, E. V. (2015). Household energy use: Applying behavioural economics to understand consumer decision-making and behaviour. Renewable and Sustainable Energy Reviews, 41, 1385-1394.

Keyword: *Socio-economic, Occupant Behavior, Energy Consumption Patterns in Residential Buildings*

Measurement of Ultrafine Particle and Black Carbon Air Pollution Exposure in Indoor and Outdoor of Residential Areas of Dhaka City

Md. Jahirul Islam *, **Niloy Dhar, Rivan Chandra Roy, Provat Kumar Saha**
Bangladesh University of Engineering and Technology, Dhaka, Bangladesh

ABSTRACT

As the level of air pollution is getting more intense day by day in Bangladesh, this issue needs to be addressed properly. Particulate matter (PM), especially PM2.5, is a significant air pollutant and it causes comparatively severe health complications. This study was aimed to measure the concentration of black carbon (BC) and ultrafine particles (UFP) – two sub-groups of PM2.5, which exhibit higher levels of hazard compared to other constituents in PM2.5, in indoor and outdoor home environments at selected residential areas located in Dhaka city. The study area encompassed 10 homes throughout the Dhaka city. Indoor exposure of BC and UFP was assessed by collecting data through 5-7 visits in each home for 30 minutes on different days during dry season. The mean indoor BC concentrations ranged from 6.5 to 10.6 $\mu g/m^3$, while outdoor BC levels varied between 7.6 and 12.8 $\mu g/m^3$. Similarly, UFP concentrations indoors ranged from 24,726 to 107,023 particles/cm^3, whereas outdoor UFP levels varied between 21,749 and 47,958 particles/cm^3. Indoor to outdoor ratios (I/O) for BC and UFP concentrations usually ranged from 0.81 to 1.05 and 0.84 to 1.43 respectively. Spatial variability of BC and UFP was analyzed further to show the trends of concentrations. Also, the analysis showed no common source of BC and UFP emissions in the study area. Surprisingly, indoor pollutant concentrations were higher in some cases, mostly due to cooking emissions. However, this analysis can further be improved by collecting data regularly.

Keyword: *Air pollution, Black carbon, Ultrafine particles*

Investigating Urban Planning Policies towards Healthy Habitat: A Case of Indian Himalayan Region

Mehul Chaudhary * [1], **Dr Venu Shree**[2]
[1] *National Institute of Technology Hamirpur*
[2] *National Institute of Technology, Hamirpur*

The rapid urbanization in developing economies presents a multifaceted challenge in creating healthy built environments that enhance occupant well-being. Addressing this challenge requires an integrated approach that merges sustainable architectural practices with effective urban planning. The rapid growth of urban populations in such regions has led to a surging demand for housing and infrastructure, often exceeding the capacity to implement sustainable practices. This imbalance results in environmental degradation, inadequate infrastructure, and health disparities among urban residents. The study investigates the effectiveness of urban planning policies in promoting healthy habitats within the Indian Himalayan Region (IHR), an ecologically sensitive area undergoing accelerated urbanization. Employing a mixed-methods approach, the research integrates spatial analysis, policy review, and stakeholder engagement to evaluate the impact of existing urban planning frameworks on public health and environmental sustainability. Key materials include spatial datasets, demographic and environmental health statistics, and relevant policy documents across various governance levels. Remote sensing and GIS tools are utilized to analyse urbanization patterns, the distribution of green spaces, and infrastructure development. The methodology combines detailed policy analysis with spatial regression to correlate urban planning features with health outcomes. Case study of a selected IHR town highlights both successful practices and critical gaps in policy execution. The findings indicate that current urban planning policies insufficiently address key health and sustainability issues, particularly in relation to resource management, and green space allocation. The study underscores the need for adaptive, context-specific urban planning approaches that integrate environmental protection, public health, and community participation.

Keyword: *Spatial Data Analysis, Policy Evaluation, Environmental Health*

Post Occupancy Evaluation: Study of Environmental Satisfaction and Work Productivity through Hot Desking System

Shreya Satodia *

Building Science, University of Southern California, Los Angeles

Shreya, Satodia University of Southern California, Los Angeles, United States of America School of Architecture, Los Angeles, United States of America satodia@usc.edu

POST OCCUPANCY EVALUATION OF LOS ANGELES OFFICE BUILDING WITH EMPHASIS ON ENVIRONMENTAL SATISFACTION AND WORK PRODUCTIVITY THROUGH HOT DESKING SYSTEM

This study presents a post-occupancy evaluation (POE) of an office building in Los Angeles, California, with an emphasis on environmental satisfaction and work productivity in a hot desking system. Hot Desking workplace arrangements are becoming popular post COVID-19 where employees follow a hybrid work culture to reduce the spread of infection. The objective of this research is to identify the factors that contribute to the environmental satisfaction and work productivity of the occupants while reviewing the IEQ performance of the selected office. In addition, this study investigates any significant environmental and physiological components that affect the user's overall IEQ satisfaction and work productivity in the hot desking system. To achieve this, this study adopted the POE methods, including survey to understand employee work comfort environment and also investigated temperature, relative humidity, CO2 levels, acoustics and illumination influencing work productivity.

The survey and IEQ parameters revealed that occupants reported significantly lower environmental satisfactions in the hot desking condition, compared to the conventional workplace environment. The finding of this research can inform the design and operation of such a novel workplace environment to improve the occupants' wellbeing and work productivity.

References:

1.10. Decision Trees — scikit-learn 1.1.3 documentation. (n.d.). Retrieved November 20, 2022, from https://scikit-learn.org/stable/modules/tree.html
73% of North Americans Do Not Consider Their Indoor Air and Water Quality Safe | Business Wire. (n.d.). Retrieved November 18, 2022, from https://www.businesswire.com/news/home/20210810005168/en/73-of-North-Americans-Do-Not-Consider-Their-Indoor-Air-and-Water-Quality-Safe.
Altomonte, S., & Schiavon, S. (2013). Occupant satisfaction in LEED and non-LEED certified buildings. Building and Environment, 68, 66–76. https://doi.org/10.1016/J.BUILDENV.2013.06.008 Asojo, A., Vo, H., & Bae, S. (2021). The Impact of Design Interventions on Occupant Satisfaction:

Keyword: *Human wellness, wellbeing, work productivity*

Study on the thermal environment in junior high school gymnasiums during evacuation

Fubuki Morihara * , **Ryouichi Kuwahara**
Yamguchi University

This study conducted simulations combining various conditions for gymnasiums in summer and winter. From the results obtained, the comfort of temperature and PMV inside gymnasiums was compared. To create a comfortable living space in an evacuation shelter, an effective air conditioning system using wall-mounted air conditioning, floor-mounted spot air conditioning, and radiant air conditioning is proposed.

The findings obtained from the simulation of a gymnasium are as follows.

It was found that the temperature differences at heights of 0.1 m, 1.1 m, and 2.1 m during summer cooling operation were low for all three air conditioning methods, creating a comfortable environment.

It was found that the PMV value at a height of 0.1 m during summer cooling operation was low for radiant air conditioning. Because radiant air conditioning is performed by spraying cold water from the floor, it is estimated that comfort immediately at the floor surface is low. During heating operation in winter, the temperature differences at heights of 0.1m, 1.1m, and 2.1m tended to be relatively low for radiant air conditioning and floor-mounted spot air conditioning, but there was a large difference in temperature above and below for wall-mounted air conditioning, and PMV also varied.
It was found that the three air conditioning methods were effective for cooling in summer, but during heating in winter, wall-mounted air conditioning alone cannot be said to provide a comfortable thermal environment, and in order to create a comfortable thermal environment, it is necessary to consider a combination of window insulation and air conditioning.

Keyword: *Thermal environment, Evacuation shelter use, Gymnasium air conditioning*

Comparative Study of the Thermal Environment of Existing Government Buildings and Its Impact on Intellectual Productivity

Imagawa Kouki * , **Ryoichi Kuwahara**
Yamaguchi University

This study compares the thermal environment of two city halls that are several decades old and their effects on intellectual productivity through a questionnaire survey. Many government buildings in Japan have been completed many years ago, and their air conditioning systems and insulation performance may have deteriorated. In addition, such deteriorated environments may adversely affect the intellectual productivity of staff. Therefore, the purpose of this study is to propose a comfortable and healthy office space by conducting a measured survey of the indoor environment and comparing two government buildings.

The results of the measurements are shown below. In both government buildings, humidity during air conditioning operation remained within the standard range of 40-70%, and there were no problems. However, temperatures of 28°C or higher were recorded at some points. There were also locations where the carbon dioxide concentration exceeded 1000 ppm and PMV (Predicted Mean Temperature) values of 1.0 or higher were also observed.

Next, the results of the questionnaire are presented. The percentages of respondents satisfied with the thermal, air, and wind environments were all less than half. This suggests that the temperature setting of air conditioning and the frequency of ventilation need to be reviewed. Regarding the impact of the indoor environment on intellectual productivity, approximately 60% of respondents felt that it "decreases.

Based on the results of the above measurements and questionnaire, it is clear that there is a causal relationship between the office environment and intellectual productivity, and that improvements are needed.

Keyword: *Thermal environment, Intellectual productivity, Renovation of government buildings*

Effects of humidity conditions in an isothermal environment on stress, arousal, fatigue, and task performance

Jin Ishii *
Meijo University

The purpose of this study is to compare and clarify the effects of high and low humidity conditions an isothermal environment on mental stress, arousal, fatigue, and task performance. The experiment was conducted in two isothermal SET environments (approximately 25°C and 27°C) under four conditions combining temperature and humidity. In this experiment, the relative humidity of the high-humidity condition was set at 70% and that of the low-humidity condition at 40%. 16 College-aged participants were recruited for the experiment. Physiological indicators of stress and arousal were measured using by EEG, ECG, and salivary cortisol levels. Fatigue was measured using a subjective survey based on a questionnaire. Skin temperature and entotic temperature were measured as physiological responses to the thermal environment. Questionnaires about psychological responses to the thermal environment were thermal sensation and thermal comfort. Task performance was evaluated by three types of tasks: visual search, simple calculation, and working memory tasks. The findings are as follows. In the isothermal SET environment, there were no significant differences in thermal comfort, stress, arousal, fatigue, or task performance due to the humidity conditions. In hot and humid regions, if the thermal environment is controlled to be the equivalent by cooling or dehumidifying in summer, the effects on thermal comfort, stress, and task performance will be the same

Keyword: *Thermal comfort, Humidity, Stress*

Assessing Insulation Needs in Residential Buildings: Model Development and Validation

Jaberswamy S * [1], **Aakash C. Rai**[1,2]
[1] *Department of Sustainable Energy Engineering, Kotak School of Sustainability, Indian Institute of Technology (IIT) Kanpur, Uttar Pradesh, 208016, India*
[2] *Chandrakanta Kesavan Center for Energy Policy and Climate Solutions, Indian Institute of Technology (IIT) Kanpur, Uttar Pradesh, 208016, India*

In India, approximately 12% of the total electricity is used for cooling residential buildings, but these buildings are usually not built in an energy-efficient manner. One effective strategy to enhance energy efficiency in such residences is incorporating insulation materials in building walls. Adding insulation increases the walls' thermal resistance (R-value), reducing heat gain and consequently lowering the building's air-conditioning (AC) energy requirements. However, to attain the desired advantages of integrating insulation, it is essential to carefully choose insulation parameters, such as the material type, R-value and placement within the wall structure (e.g., on the exterior or interior), by considering the local climate. An improper application can even increase the buildings' AC needs and degrade indoor comfort levels; therefore, assessing insulation needs for buildings is necessary.

This study involves year-round monitoring of AC energy consumption, indoor temperature, and relative humidity in individual rooms of a single house in a residential apartment. An energy simulation model was developed using EnergyPlus, and the model's accuracy was validated against the monitored hourly data using established statistical error indicators, specifically normalised mean bias error (NMBE) and coefficient of variation of the root mean square error (CVRMSE). The results showed that the model effectively captured the observed trends and reasonably matched the actual data, with NMBE and CVRMSE values typically within ±4% and ±24%. The validated model can assist in developing design guidelines for insulation in Indian residences and provide a cost-effective solution for reducing energy use and carbon footprints in the Indian building sector.

Keyword: *Building energy simulation, Envelope heat transfer, Insulation design guidelines*

Perceived Productivity, Discomfort, and Sensory Interactions: Evaluating the Combined Effects of IAQ, Acoustic, and Visual Environments

Tanmay Dave * **, Rune Korsholm Andersen, Pawel Wargocki**
Technical University of Denmark

Modern humans spend most of their lives indoors, dividing their time among three major activities – leisure (home), work (office/school), and transport (vehicles). As awareness of work-life balance grows, creating work environments that promote human productivity has become increasingly important. While past research has extensively studied the effects of indoor environmental quality (IEQ) on productivity, it often treats each parameter of IEQ in isolation, overlooking the multi-modal nature of human perception. This study aims to advance understanding of the multi-modal effects of IEQ and how occupant preferences shift in states of concurrent discomfort.

We conducted an experiment with 30 participants exposed to eight distinct IEQ conditions, combining factors related to indoor air quality (IAQ), noise, and glare in a full-factorial, within-subject design. The study included two levels for each parameter: speech (quiet or babble noise), glare (absent or present), and air quality (low ventilation with pollution or high ventilation without pollution), reflecting the three IEQ factors considered. Participants evaluated the IEQ conditions based on perception, comfort, acceptability, and preferences regarding factors that enhanced or hindered their productivity. Environmental variables were monitored and controlled to isolate the effects.

The findings are expected to deepen our understanding of how IEQ factors interact and provide insight into the hierarchy of discomfort experienced by occupants. These results will offer actionable strategies for designing workspaces that promote productivity and provide valuable implications for office design.

Keyword: *Multi-perceptual, Human productivity, Comfort*

Indoor Air Quality Monitoring and Health Risk Assessment in Gyms

Amrutha Prabhakar * , **ASWATHY E.V.**
National Institute of Technology Calicut

Gyms are unique indoor environments where high physical activity levels exacerbate the generation and dispersion of fine particles, potentially affecting air quality and human health. Indoor air quality (IAQ) in air-conditioned gyms plays a critical role in the health and performance of occupants - gym-goers, who typically engage in high-intensity activities for 1–2 hours per session daily, face elevated exposure to airborne pollutants due to increased respiratory rates, potentially leading to respiratory issues and other health risks. This study investigates IAQ in gyms, focusing on particulate matter (PM_{10}, $PM_{2.5}$, PM_1), carbon dioxide (CO_2), alongside other comfort parameters such as temperature and humidity.

PM and CO_2 was measured using real-time monitoring equipment (Grimm MiniLAS and Delta Ohm IAQ Meter), and data collection was done across three consecutive weekdays (Tuesday, Wednesday, Thursday) encompassing peak occupancy hours (morning and evening) and non-working hours. PM and CO_2 data was logged at one-minute intervals, and the results were analysed using descriptive and inferential statistical methods, including Pearson's correlation to examine relationships between pollutants and IAQ parameters.

The observed PM and CO_2 concentrations in gyms were exceeding the WHO (PM_{10} : 45 μg/m^3 $PM_{2.5}$: 15 μg/m^3) and ASHRAE (CO_2 : 1000 ppm) recommended limits respectively. The findings aim to assess potential health risks for gym-goers associated with PM and CO_2. Recommendations for mitigation strategies are expected to guide policymakers and gym operators in fostering healthier gym indoor environments.

References:

Peixoto, C., Pereira, M. do C., Morais, S., & Slezakova, K. (2023). Assessment of indoor air quality in health clubs: insights into (ultra)fine and coarse particles and gaseous pollutants. Frontiers in Public Health, 11(November), 1–14. https://doi.org/10.3389/fpubh.2023.1310215

Parhizkar, H., Taddei, P., Weziak-bialowolska, D., Mcneely, E., Spengler, D., Guillermo, J., & Laurent, C. (2023). Effect of ventilation strategies and air purifiers on the children's exposure to airborne particles and gaseous pollutants in school gyms. Building and Environment, 110984. https://doi.org/10.1016/j.buildenv.2023.110984

Andrade, A., Dominski, F. H., Pereira, M. L., de Liz, C. M., & Buonanno, G. (2018). Infection risk in gyms during physical exercise. Environmental Science and Pollution Research, 25(20), 19675–19686. https://doi.org/10.1007/s11356-018-1822-8

Keyword: *Particulate matter, Carbon dioxide, Indoor air, Inhalation Health risk*

Assessing Respiratory Health Impacts of Traditional Fuels in different Built Environments in Northwest Rajasthan

Harshita Karri * , **Dr Rajyalakshmi Garaga**
Centre for Emerging Technologies and Sustainable Development, Indian Institute of Technology Jodhpur, Rajasthan, India- 342037

ABSTRACT

Indoor air pollution, primarily caused by the combustion of polluting fuels like wood, animal dung, and agricultural residues, is a critical global health issue, contributing to approximately 3 million premature deaths annually (WHO). This study surveyed 330 households across Rajasthan's arid Thar Desert region, including Jodhpur, Jaisalmer, Barmer, Bikaner, Nagaur, Churu, and Hanumangarh, to analyze the impacts of housing types and cooking fuels on respiratory health. Results showed that Dung houses (Kutcha) predominantly used conventional fuels such as solid biomass fuels, while Dung+Stone (Semi-Pucca) and Stone (Pucca) houses increasingly adopted cleaner fuels like LPG. The data highlighted a significant correlation between biomass fuel use and respiratory diseases, including asthma and tuberculosis. Among Dung-only households, 46.43% reported respiratory symptoms compared to 42.86% in Dung+Stone households and 10.71% in Stone-only households. Poor ventilation and prolonged exposure to indoor air pollution were key contributors to these disparities. Cleaner fuel adoption was higher in Dung+Stone (34.74%) and Stone only (31.16%) households compared to Dung only (21.65%) households, indicating a gradual shift toward cleaner fuels in semi-modern homes. This study emphasizes the immediate requirement for interventions, such as improving ventilation in traditional homes, promoting cleaner fuels, and addressing socio-economic barriers. Future research should focus on developing sustainable intervention plans, enhancing traditional chimney designs, and exploring alternative energy sources and education initiatives to mitigate health risks and improve rural living conditions.

Keyword: *Indoor air pollution, Respiratory health, Conventional fuels*

Industrial Exposure Driven Risk Mapping for Industrial Neighbourhoods: A Case Study of Subash Nagar, Jeedimetla, Hyderabad

Sandeep Budde * [1], **Sandeep Agrawal**[2], **P S Chani**[3]
[1] *Univeristy of Alberta, Canada- IIT Roorkee, India*
[2] *Urban Environment Observatory, School of Urban and Regional Planning, University of Alberta, Edmonton AB - T6G 2R3, Canada*
[3] *Built Environment Lab, Department of Architecture and Planning, Indian Institute of Technology, Roorkee - 247667, India*

Industrial zones in urban areas are hubs of economic activity but often pose significant risks to adjacent residential neighborhoods due to emissions. This study focuses on Subash Nagar in Jeedimetla, Hyderabad, to develop a risk mapping framework driven by industrial exposure. The study incorporated data from 529 buildings and 20 building clusters, categorized based on building volume and total opening area, to assess exposure dynamics under varying building typological conditions. Key parameters such as net floor area, gross floor area, building volume, total window and door areas, wall window ratio, and compactness were analyzed to identify building typologies and their correlation with risk scores. Parameters like total window openings, wall window area, and building orientation were critical for evaluating pollutant dispersion and accumulation in low-pressure zones.

Risk zones were identified through a CFD simulation method and analysis of percentage low-pressure areas, integrating environmental data, building density, and proximity to industrial units. The analysis revealed significant pollutant hotspots within built-up areas in different building clusters, with larger buildings containing more partitions experiencing amplified low-pressure zones. The influence of low-pressure regions and low wind velocity was particularly evident in pollutant stagnation, increasing risk exposure for specific clusters with more than three partitions within a building plan.

This case study underscores the importance of exposure-driven risk mapping in addressing environmental challenges in industrial neighborhoods. The proposed methodology demonstrates potential replication across other settlements in industrial clusters, contributing to urban resilience.

References:

Amirebrahimi, S., Rajabifard, A., Mendis, P., & Ngo, T. (2016). A framework for a microscale flood damage assessment and visualization for a building using BIM–GIS integration. International Journal of Digital Earth, 9(4), 363–386. https://doi.org/10.1080/17538947.2015.1034201

Bertsch, R., Glenis, V., & Kilsby, C. (2022). Building level flood exposure analysis using a hydrodynamic model. Environmental Modelling and Software, 156(August), 105490. https://doi.org/10.1016/j.envsoft.2022.105490

Blanco-Vogt, Á., Haala, N., & Schanze, J. (2015). Building parameters extraction from remote-sensing data and GIS analysis for the derivation of a building taxonomy of settlements – a contribution to flood building susceptibility assessment. International Journal of Image and Data Fusion, 6(1), 22–41. https://doi.org/10.1080/19479832.2014.926296

Keyword: *Industrial Exposure, Risk Mapping, Low Pressure Zones*

An AI-ML based Decision Support Tool for Optimizing Green Infrastructure to Mitigate Air Pollution in Urban Built Environments

Cherith Reddy Yerabolu, Geethika Choudhary Yadlapalli, Monisha Kollipara, Sri Dhruti Tikkisetti, Surekha Tetali *

Mahindra University

One critical impact of climate change, driven by rising greenhouse gas emissions, is its adverse effect on air quality, which directly impacts human health and well-being. While addressing primary sources, such as emissions from energy production and transitioning to electric vehicles, is crucial, decentralized solutions within the urban built environment offer significant potential to mitigate air pollution and enhance resilience. Research has highlighted the effectiveness of nature-based solutions, particularly green infrastructure, in reducing air pollution. However, the efficiency of such interventions often varies based on building designs, systems, and contextual factors. This study addresses this gap by integrating insights from existing research and contextual data to optimize green infrastructure for air pollution reduction in and around buildings. A comprehensive database will categorize building systems and green infrastructure strategies based on their pollution reduction potential. Using text-mining pipelines with Natural Language Processing (NLP) and Machine Learning (ML), the database will synthesize data from peer-reviewed literature and product specifications to systematically catalog building systems and green infrastructure solutions. By integrating these datasets with localized air quality data, a multi-step optimization process will leverage ML and reinforcement learning to identify the optimal combination of building systems and green infrastructure for achieving the lowest possible air quality index. This decision support tool will empower homeowners, builders, architects, urban planners, and policymakers to make data-driven decisions, advancing sustainable urban development and improving indoor and outdoor air quality, fostering healthier and more resilient communities.

Keyword: *Green Infrastructure, Optimization, Air Pollution*

Does radon have health effects other than lung cancer? A systematic review and meta-analysis

<u>Corinne Mandin</u> * , Afi Mawulawoe Sylvie Henyoh, Olivier Laurent, Enora Clero
Nuclear Safety and Radiation Protection Authority, France

Radon is a major indoor air pollutant. To date, lung cancer is the only well-established health effect associated with radon exposure. To summarize evidence on other potential health effects of radon exposure, we performed a comprehensive qualitative and quantitative synthesis of the available literature on radon exposure and health effects other than lung cancer.

Studies published from January 1990 to March 2023 were identified in PubMed, ScienceDirect, Scopus, ScieLo, and HAL. We estimated average weighted standardized incidence ratios (metaSIR), standardized mortality ratios (metaSMR), and risk ratio (metaRR) per 100 Bq/m^3 increase in radon concentration by combining estimates from the eligible studies using the random-effect inverse variance method.

A total of 129 studies were included in the systematic review and 40 studies in the meta-analysis. The results of the meta-analyses showed no statistically significant association. However, the estimated exposure-risk associations were positive and close to the statistical significance threshold for: lymphohematological cancer incidence in children (metaRR = 1.01; 95%CI: 1.00–1.03; p = 0.08); malignant melanoma mortality among adults in the general population (metaRR = 1.10; 95%CI: 0.99–1.21; p = 0.07); liver cancer mortality among mine workers (metaRR = 1.04; 95%CI: 1.00–1.10; p = 0.06); intestine and rectal cancer mortality combined among mine workers (metaRR = 1.02; 95%CI: 1.00–1.04; p = 0.06).

While carcinogenic effects other than lung cancer and non-carcinogenic effects associated with radon exposure are biologically plausible, the results of this meta-analysis did not show radon-related risks other than lung cancer based on available epidemiological studies.

Keyword: *Radon, Meta-analysis, Epidemiology*

Effective Filtration Media for Healthy Indoor Air for Residential and Office Spaces

Manu Shetty *
Hollingsworth and Vose, Mysore,India

Abstract submission for Healthy Buildings 2025
Submitter: Hollingsworth & Vose, Co.

Format: Poster
Theme: Effective Filtration Media for Healthy Indoor Air for Residential and Office Spaces

Quality of indoor air plays a major role in human health and productivity. Humans spend over 90% of their time on average indoors, and without effective filtration and ventilation, the air inside can be 2-5 times, even up to 100 times more polluted than outdoors.

The science of filtration can seem complex and confusing to those outside the industry. A common misperception is that filters are "commodity items" without significant differentiation from one to another; however, the heart of a filter - the performance layer, is made up of filtration media which can vary greatly depending on the materials and technology used to create it and the most effective filters feature the most effective filtration media.

We propose creating a poster that will highlight the important societal need for cleaner air in homes and office buildings and need for filters with the most effective filtration media for healthier air. This can also benefit building owners and facilities managers by providing healthier air, less energy from the HVAC system, lower energy bills, and even less outdoor air pollution by reducing scope 2 GHG emissions.

Summary: We aim to support the education of stake holders on how to select filters with the most effective filtration media to create the best wellness, cost of ownership, and energy outcomes for the healthy building landscape.

Keyword: *Filtration media*

Microclimatic Variations and Indoor Thermal Comfort in a Residential Building During Heat Wave and Typical Summer Conditions in Nashik

Ketaki Joshi *, **Prashant Anand**
Department of Architecture and Regional Planning, Indian Institute of Technology Kharagpur

The increasing frequency of extreme heat events, exacerbated by climate change, has raised concerns regarding indoor thermal comfort in residential buildings, particularly in developing cities like Nashik, India. This study investigates the impact of outdoor microclimatic variations on indoor thermal comfort in the bedroom of a top-floor flat in a 10-storey apartment during a heat wave day (HWD) and a typical summer day (TSD). Outdoor and indoor air temperature (Ta) and relative humidity (RH) were simultaneously recorded at 5-minute intervals. HWD was defined based on official criteria set by Indian Meteorological Department, while TSD was identified based on historical climatological norms. The analysis was done using the Discomfort Index (DI) and Cooling Degree Hours (CDH) during usage hours, daytime (1:00 PM to 4:00 PM) and night-time (9:00 PM to 9:00 AM), under the operation of mixed-mode ventilation systems.

The analysis of the DI on a HWD, based on Thom's classification, showed severe discomfort at night, peaking at 29.33°C at 9:00 PM and remaining above 28°C until early morning. Daytime DI ranged from 27.10°C to 27.90°C, indicating high discomfort during the HWD, even with cooling systems operative. Total CDH increased by 59.87% on the HWD, with a notable increase in night-time CDH by 111.26% and a moderate rise of 19.54% during daytime usage hours compared to TSD.

This evidence-based study, grounded in onsite measurements, emphasizes the importance of resilient building designs and energy-efficient cooling strategies to enhance Indoor Environmental Quality (IEQ) and thermal resilience in the face of escalating climate challenges.

Keyword: *Discomfort Index, cooling energy demand, heat wave resilience*

IAQ resilience indicators

Jelle Laverge *

Ghent University

In their 2023 paper on 'the resilience score metric', Al Assaad et al. (1) proposed a metric for the assessment of the resilient response of a building to a disturbance or shock in terms of the indoor air quality (IAQ). The metrics they describe, absorbtivity, recovery and shock impact, are derived from existing metrics for thermal shocks and provide a good image of the resulting IAQ, but provide little information on how the system will react to other shocks and are difficult to relate to HVAC or IAQ management system design requirements. In this paper, we analyze the requirements for resilience indicators from a risk mitigation perspective on the added value brought by an IAQ management system and propose an additional set of indicators to characterize both the shock and the response that are based on intrinsic properties of the system such as the estimated frequency of occurrence, the capacity of the system to reduce the pollutant concentration and the time constant of the system. With these indicators, the designer and their client can make informed decisions on the design specifications and the desired level of resilience.

References:

https://doi.org/10.1016/j.buildenv.2023.110669

Keyword: *Resilience, IAQ, Risk Analysis*

Behavioral and Socio-Economic Dimensions of Indoor Air Quality: Insights into Public Perceptions and Interventions in India

Ritwik Agarwal[1], **Vishal Garg**[1,2], **Jyotirmay Mathur**[3], **Prakarsh Singh**[2], **Aviruch Bhatia**[2]

[1]*International Institute of Information Technology, Hyderabad, India*

[2]*Plaksha University, Mohali, Punjab, India*

[3]*Malaviya National Institute of Technology Jaipur, India*

Air pollution is a growing concern in India, specifically in the National Capital Region (NCR) which is one of the world's most polluted regions. However, there is limited research on how people understand air pollution, the strategies they use to reduce its impact on their well-being, and the effectiveness of these methods. This study addresses these gaps by examining public awareness of air pollution in India and the effectiveness of solutions used to improve Indoor Air Quality (IAQ).

As part of the study, ethnographic in-person interviews were conducted in India's national capital region to gather detailed socio-economic perspectives. Results indicate that while awareness of the Air Quality Index (AQI) is widespread (88% of total respondents), understanding of PM2.5 and its health effects remain limited (33% of respondents). Younger, educated residents are more likely to consider relocating to lower-AQI southern Indian cities if offered comparable job opportunities, reflecting migration as a mitigation strategy. Air purifiers were identified as a common intervention, but participants cited effectiveness of purifiers (40% of those who don't own the device) and limited awareness of IAQ as significant barriers. Most people who use air purifiers report perceived health benefits but do not consider them as a permanent solution.

These observations highlight the need for better information on solutions such as air purifiers and other affordable strategies to manage IAQ. Targeted, localized awareness campaigns on PM2.5 and mitigation programs could also be employed by governing bodies.

Keywords: *Air Quality Awareness, Air Purifiers, Preventative Actions*

Behavioural and Socio-Economic Dimensions of Indoor Air Quality: Insights into Public Perceptions and Interventions in India

Ritwik Agarwal[1]*, Vishal Garg[12], Jyotirmay Mathur[3], Prakarsh Singh[4], Aviruch Bhatia[2]

[1] Center for IT in Building Science, International Institute of Information Technology, Hyderabad, India
[2] Indorama Ventures Center for Clean Energy, Plaksha University, Mohali, India
[3] Center for Energy and Environment, Malaviya National Institute of Technology, Jaipur, India
[4] Data Science, Economics & Business (DSEB), Plaksha University, Mohali, India
*Corresponding email: aritwik93@gmail.com

SUMMARY
Air pollution is a major concern in India's National Capital Region (NCR), one of the world's most polluted areas. This study explores public awareness of air pollution and the perceived effectiveness of strategies to improve Indoor Air Quality (IAQ). Insights were gathered through in-person interviews across Delhi-NCR, representing a range of socio-economic backgrounds. Findings reveal widespread awareness of the Air Quality Index (88% of total respondents), but limited understanding of $PM_{2.5}$ and its health risks (33%). Younger, educated residents are more likely to consider relocating to less polluted cities if comparable opportunities exist, highlighting migration as an adaptation strategy due to air quality concerns. Air purifiers are commonly used, though limited awareness of IAQ and effectiveness of purifiers present significant barriers (40% of non-purifier owners). These findings highlight the need for better public information on $PM_{2.5}$ and affordable indoor air quality solutions.

KEYWORDS
Air Quality Awareness; Air Purifiers; Migration Intention; Preventive Attitude

1 INTRODUCTION
Air pollution is one of the leading global public health threats, responsible for an estimated seven million premature deaths annually (WHO, 2025). Countries undergoing rapid urbanization and industrial growth, such as India and China, face heightened risks due to deteriorating air quality (Dobbs & Sankhe, 2010). In India, the problem is especially severe in the Delhi National Capital Region (NCR), where winter pollution episodes are intensified by adverse weather conditions, vehicular emissions, industrial activities, biomass burning, and construction dust (Khanna & Sharma, 2020).

Despite a range of policy interventions, exposure to hazardous air pollutants—especially $PM_{2.5}$—remains critically high. (Greenstone et al., 2021) found that indoor $PM_{2.5}$ levels in Delhi households exceeded WHO limits by over 20 times, with minimal to no differences between higher and lower-income households. Even when provided with indoor air quality monitors (IAQMs) and subsidized air purifiers, many households were reluctant to adopt mitigation measures, citing cost concerns and doubts about their effectiveness.

Although attention is often focused on outdoor air quality, IAQ—which is closely influenced by outdoor pollution—remains an under-addressed issue. (Gupta et al., 2024) demonstrated that even green-rated homes in Delhi exhibited high $PM_{2.5}$ exposures, underscoring the urgent need for large-scale monitoring of IAQ and its links to occupant behaviours.

This study investigates perceptions and behaviours related to air pollution among residents of Delhi-NCR. Through interviews, it explores awareness of air quality indicators, strategies used to mitigate exposure, and barriers to adopting IAQ interventions such as air purifiers. The findings offer insights to inform targeted awareness initiatives and promote effective IAQ management strategies, while highlighting the socio-economic costs of poor air quality in the region.

2 METHODS

Data for this study was collected through ethnographic, in-person interviews with 101 households across Delhi-NCR, conducted over February 2025. The four cities selected—Delhi, Gurgaon, Noida, and Ghaziabad—offered a diverse representation of socio-economic groups, from low-income communities to affluent urban residents. Participants included both air purifier users and non-users to ensure balanced insights. Each interview, lasting approximately 25 minutes, explored perceptions of air quality, mitigation behaviours, and IAQ management, supplemented by randomized spot IAQ assessments. The interview design focused on three key areas: Awareness, Mitigation practices, and use of air purifiers, while also gathering demographic data to analyse socio-economic correlations. All participants provided informed consent, and all ethical guidelines were followed while conducting the study. As seen from Table 1, all participants were interviewed using a prepared structure with open-ended questions to capture qualitative responses.

Table 1: Interview Structure used to collect household responses

Category	Response Type	Objective
Awareness & Perception	Word associations, word clouds, Likert scale	Assess knowledge of AQI/$PM_{2.5}$, Pollution sources, and satisfaction levels.
Preventive Strategies	Multiple-choice, subjective responses, Yes/No	Examine mitigation measures, migration intentions, and barriers to action.
Air Purifier Adoption & Usage	Choice questions with reasoning, explanations	Analyse triggers, usage patterns, and satisfaction with air purifiers.
Behavioural Impact of IAQ Monitoring	Yes/No with reasoning	Assess how IAQ monitoring influences purchase decisions and behaviour change.

2.1 Measures of Air Quality Awareness

To evaluate air quality and $PM_{2.5}$ awareness among participants, a structured set of interview questions was administered, designed to capture both awareness and knowledge-based behaviours. The assessment incorporated both binary responses (e.g., yes/no) and a composite knowledge score. Participants were asked two direct awareness questions: (1) "Do you know what AQI is?", and (2) "What do AQI and $PM_{2.5}$ represent?"

Responses such as "no," "don't know," "no idea" or similar uninformative answers were coded as uninformed. In addition to these binary measures, knowledge score out of 5 was calculated for each respondent based on the following criteria:

Table 2: Mapping Knowledge score to Air Quality Awareness

Component of Knowledge score	Criteria for 1 Point Awarded
Awareness of AQI	Correctly identifies or recognizes what AQI is
Awareness of local $PM_{2.5}$ sources	Able to name or describe nearby sources of $PM_{2.5}$
Frequency and method of checking AQI	Reports checking AQI daily or weekly and indicates the method
Awareness of $PM_{2.5}$ as a pollutant	Indicates knowledge of $PM_{2.5}$ as a harmful component of pollution
Changed behaviour because of Air Quality	Reports taking precautions (e.g., staying indoors, mask use)

3 RESULTS AND DISCUSSION

The results highlight three key insights into residents' awareness, behaviours, and decision-making in response to air pollution. The key findings from this phase are as follows:

Awareness of AQI and health risks

The results indicate that air purifier ownership is associated with higher levels of environmental health awareness. Specifically, for Air Quality Index (AQI) awareness, air purifier owners reported a mean awareness score of 0.974 (97.4% aware), compared to 0.823 (82.3%) among non-owners. An independent samples t-test revealed a statistically significant difference between the two groups, $t(df) = 2.720, p = 0.0078$, suggesting that this difference is unlikely to be due to random variation. For $PM_{2.5}$ awareness, the disparity was more pronounced: owners reported a mean awareness score of 0.579 (57.9% aware) while non-owners reported 0.210 (21.0% aware). The t-test yielded $t(df) = 3.828, p = 0.00029$, indicating a highly significant difference.

When comparing the magnitude of these effects, the higher t-value and lower p-value associated with $PM_{2.5}$ awareness suggest a stronger relationship between air purifier ownership and PM2.5 awareness than with AQI awareness. In other words, the gap in $PM_{2.5}$ awareness between owners and non-owners is more substantial than the gap in AQI awareness. However, while analysing the overall sample, it is found that while basic AQI knowledge is common, a detailed understanding of $PM_{2.5}$ is limited among all residents. Overall, 88% of residents were aware of AQI, while only 33% demonstrated clear understanding of $PM_{2.5}$

Furthermore, a knowledge score/rating was assigned to each household as described in the methods section above. As seen from Figure 1, air purifier ownership is linked to a statistically significant rise in knowledge ratings ($\beta \approx 1.04$). This suggests that owners tend to score about one point higher, on average, than non-owners and are thus more aware of Air Quality around them.

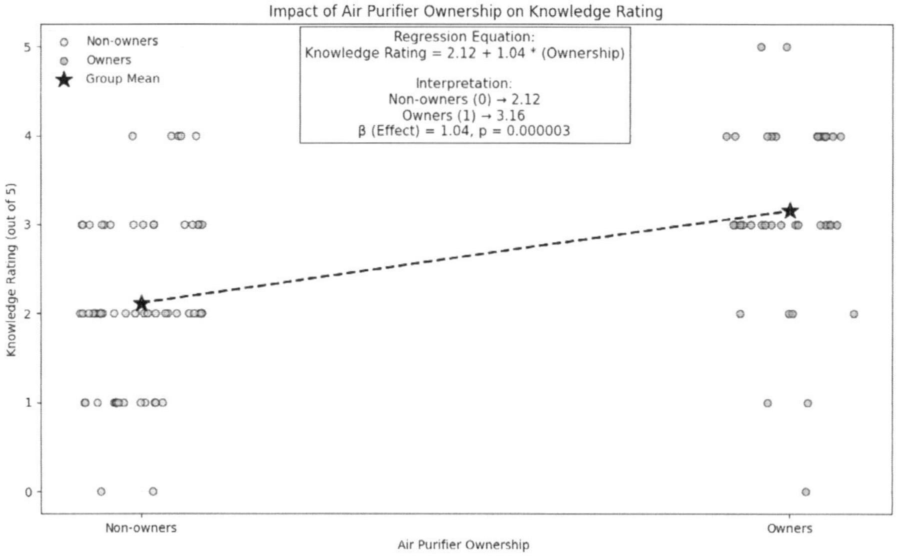

Figure 1 : Distribution of Air Quality Knowledge ratings

The qualitative differences in what people are saying further underscore this gap. Those who have been assigned higher knowledge rating described the sources of $PM_{2.5}$ with greater clarity. In contrast, many respondents who did not own air purifiers used very brief statements or expressed uncertainty about the significance of $PM_{2.5}$ (e.g., "no idea"). Overall, the average score in both groups is not significantly high and suggests that efforts to improve public education on air quality—especially detailed information on $PM_{2.5}$, its sources, and its health impacts—could be beneficial.

Air Purifier Perception

A significant proportion of households reported using air filtration methods to improve their indoor air quality. Approximately 30% mentioned employing some form of air filtration—this includes indoor air purifiers, car air purifiers, and air conditioning systems with circulation filters. Among air purifier users, 63.16% use them daily while 36.84% do not, with daily users showing 100% AQI awareness and 58.33% $PM_{2.5}$ awareness versus 92.86% and 57.14% respectively for non-daily users. Despite these higher percentages, the t-test results ($p = 0.33556$ for AQI and $p = 0.945$ for $PM_{2.5}$) indicate that these differences are not statistically significant. This means that although daily users seem to be marginally more aware, the effect of daily usage on AQI and $PM_{2.5}$ awareness levels is not strong enough to be considered meaningful.

Instead, this difference in usage frequency is fuelled by effectiveness of air purifiers and if owners felt that regular usage is giving them some sort of benefits. Among air purifier owners, 75% perceived some form of benefit. Of these, 30% described specific health improvements, including reduced respiratory symptoms and cleaner indoor air. However more than 45% reported perceived or psychological benefits, such as feeling better or more comfortable indoors. Despite these reported advantages, 40% of owners expressed that they viewed purifiers as a temporary or limited solution, primarily due to their indoor-only function and inability to address outdoor air pollution.

For non-owners, several factors contributed to non-adoption. The majority, 40% of non-owners, believe their household's indoor air quality is already sufficiently good or express doubts about the effectiveness of air purifiers, particularly in addressing outdoor air pollution. Concerns about potential harm from prolonged use—such as weakening the immune system or forming an unwanted dependency or a day-to-day habit—account for 30% of responses. Additionally, 20% reported they had never considered purchasing an air purifier, while 10% cited cost as the primary barrier to adoption.

These findings indicate that, while a significant proportion of owners perceive benefits from air purifier use, most reported improvements are subjective. Only a smaller subset reports clear, measurable health outcomes. Additionally, perceived limitations, IAQ awareness and scepticism about effectiveness continue to influence public decisions around home interventions for tackling poor air quality, with socio-economic factors (Figure 2) playing a significant role in both adoption of air purifiers and awareness around $PM_{2.5}$.

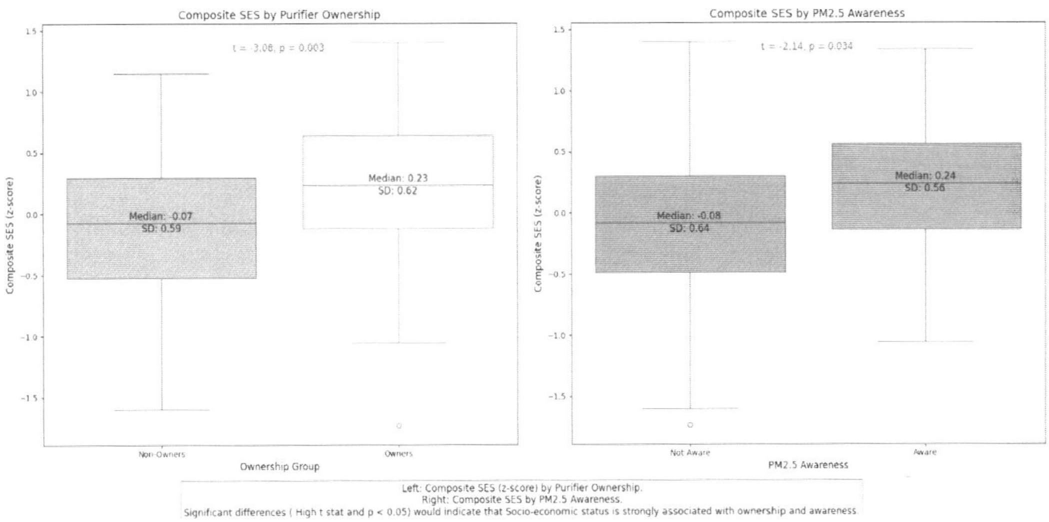

Figure 2: Education Level, Household size and Income influence Air purifier purchase decisions

Migration Intention

Respondents were also asked about behavioural changes due to poor air quality and the precautions they take to reduce the effect of pollution on them. While Mask usage is a common mitigation measure, more than 40% respondents who were unaware of AQI report taking no precautions at all, highlighting a significant gap in public awareness. Additionally, households were asked about their willingness to consider migration as a response to the persistently poor AQI levels in the Delhi-NCR region. Logistic regression shows that there is a significant age effect. Each additional year of age was associated with about a *4% decrease* in the odds of planning to migrate (OR 0.96, 95% CI 0.92–1.00, p = 0.009). In other words, younger respondents showed significantly higher willingness to migrate than older respondents. Among the under-30 group, 70% intended to migrate, compared to just 23% of those over 50. Gender differences were minimal: 53% of women and 57% of men intended to move, but this was not statistically significant. Similarly, ownership of Air Purifiers did not corelate to migration intent as both groups were equally likely to move away from Delhi. Thematic analysis revealed health concerns as the primary driver for moving, while family obligations and job security were key barriers. When looking at destination preferences, women were more inclined to move to rural areas and hill stations, often prioritizing cleaner air for their families. Men, on the other hand, tended to favour southern Indian cities like Bangalore and Hyderabad, typically for career opportunities, or opted to relocate within the Delhi-NCR region. Older respondents (average age 48) preferred relocating to rural villages or hill stations. Younger individuals in their 20s and 30s mostly chose major cities like Bangalore or Mumbai or planned to move abroad.

4 CONCLUSIONS

The fact that Delhi-NCR continues to experience hazardous air quality levels, and that a significant portion of the population remains inadequately informed about $PM_{2.5}$, makes raising public awareness and promoting effective IAQ interventions a key challenge. The findings from this study clearly show that improving public knowledge about $PM_{2.5}$ and its health impacts, providing access to affordable IAQ solutions, and encouraging preventive behaviours are all important motivational factors in driving positive behavioural change. The data highlights key directions for future interventions:

- Enhance public education campaigns that emphasize the health risks of $PM_{2.5}$, particularly tailored to non-users of air purifiers and vulnerable populations.
- Promote accessible and affordable IAQ technological interventions, addressing both cost barriers and misconceptions regarding their efficacy.
- Encourage regular IAQ monitoring and its integration into household decision-making, as knowledge is closely linked with precautionary measures.
- Consider retention initiatives and investments in green infrastructure to improve urban liveability. This could help discourage relocation, particularly as younger, educated residents are more inclined to migrate for better air quality.

Improving IAQ in Delhi-NCR requires more than temporary solutions like air purifiers, which many residents view as inadequate or unsustainable. Most people express a preference for permanent, long-term interventions that address the root causes of poor air quality. Government and policy initiatives should be aimed at improving long-term indoor air quality and overall well-being which are just as important as short-term solutions, particularly for residents considering major life changes such as relocating or investing in healthier living spaces. Insights related to government trust, willingness to pay, and behaviours surrounding air purifier usage will be explored in greater detail in future work.

5 ACKNOWLEDGEMENTS

The research presented in this paper was supported by Indorama Ventures Center for Clean Energy, Plaksha University.

6 REFERENCES

Dobbs, R., & Sankhe, S. (2010). Comparing urbanization in China and India. *McKinsey Quarterly*.

Greenstone, M., Lee, K., & Sahai, H. (2021). *Indoor Air Quality, Information, and Socio-Economic Status: Evidence from Delhi*. https://ssrn.com/abstract=3767009

Gupta, R., Zhao, Y., Garg, V., & Mathur, J. (2024). *Empirical Examination of Trends in Indoor Air Quality in a Sample of Urban Indian Residences*. 134–143. https://doi.org/10.62576/ojim5461

Khanna, I., & Sharma, S. (2020). Could the National Capital Region serve as a control region for effective air quality management in Delhi? *Policy Brief, CCAPC/2020/02, Collaborative Clean Air Policy Centre, New Delhi*.

WHO. (2025). *Air Pollution impact*. https://www.who.int/health-topics/air-pollution#tab=tab_2

Adaptive Thermal Comfort and Smart Building Management: Case Studies of HVAC Optimization and IBMS Integration in Warm and Humid Climates

Ketaki Gupte * [1], **Dr. Deva Dubey**[2], **Kundan Bharambe**[1], **Deepali Hegde**[3]
[1] *Research Scholar – RICS SBE, Amity University, Mumbai, India*
[2] *Associate Professor – RICS SBE, Amity University, Mumbai, India*
[3] *Deputy General Manager (ESG & Sustainability) - Godrej Properties Ltd, Mumbai, India*

India's commitment to net-zero carbon emissions by 2070 declared at COP26, has emphasized the need to prioritize energy efficiency for achieving sustainable, carbon neutral built environments. Metropolitan cities like Mumbai, have a congregation of numerous corporate office buildings catering various multinational firms that depend on HVAC systems to achieve Thermal Comfort. This comes at the cost of excessive energy demand, accounting for nearly 60% of building's total energy consumption. Despite substantial energy consumption, most of the HVAC systems operate below optimal efficiency. This results in energy wastage and increased operational costs. Intelligent Building Energy Management Systems (IBMS) have the potential to achieve energy efficiency of up to 30% by optimizing both building operations and performance of HVAC systems. However, their effectiveness in the warm and humid climate of Mumbai remains underexplored. Additionally, the diverse occupancy patterns and occupant preferences in these offices pose a challenge to the HVAC optimization.

This study examines the impact of IBMS optimised HVAC systems on energy efficiency and occupant thermal comfort by conducting a comparative analysis to evaluate a couple of few IBMS-integrated office buildings in Mumbai, using annual real-time IBMS data that monitors temperature, humidity, CO_2 levels, and energy consumption in conjunction with questionnaire-based occupant surveys. Thermal comfort in these offices is evaluated against the ASHRAE 55, IMAC, and WELL Building Standards. Energy performance is compared with ECBC 2017 and ISHRAE IAQ guidelines.

Initial findings show that IBMS optimizes HVAC systems performance by reducing energy consumption and maintaining better indoor air quality. However, occupants' thermal comfort varies based on occupancy patterns and their perception of built environment. This study offers insights for developing a framework for energy-efficient, user-centric office design, which could foster technological advancements in smart building management, especially in warm and humid climates, supporting SDG 7, SDG 11, and SDG 13.

References:

ASHRAE. (2017). ANSI/ASHRAE Standard 55-2017: Thermal Environmental Conditions for Human Occupancy. American Society of Heating, Refrigerating and Air-Conditioning Engineers.

Indian Society of Heating, Refrigerating and Air Conditioning Engineers (ISHRAE). (2019). Indoor Air Quality Standard – ISHRAE IAQ Standard: 2019. ISHRAE Technical Committee.

INVESTIGATING INDOOR THERMAL COMFORT OF APARTMENTS THROUGH THE SYNERGY OF VERTICAL GREENERY SYSTEMS AND JAALI IN A WARM HUMID CLIMATE: A CASE OF TRIVANDRUM

Sunu Sam *, **Dr. Deepa Rani**
College of engineering Trivandrum

The study investigates the integration of Vertical Greenery Systems (VGS) with jaali to enhance indoor thermal comfort in warm, humid climates, Trivandrum. While VGS, categorized as green facades and living walls, are increasingly used for their aesthetical, economic, environmental, and social benefits, limited research exists on their combination with traditional jaali systems. This research compares thermal comfort of a green facade system, featuring Yellow Trumpet Vine, and its combination with clay cones against a baseline condition through field experiments and simulations. Indoor thermal comfort parameters such as air temperature, relative humidity, mean radiant temperature, and air velocity were measured for 15-minute interval alongside baseline conditions for span of 16 days for east orientation and 8 days for south orientation, at second floor level of an apartment. For the east orientation, the green façade reduced temperatures by 1.4°C and the combination façade by 3.3°C, after accounting for a 2°C reduction due to room volume and balcony size. For the south orientation, the green façade reduced temperatures by 1°C and the combination façade by 1.5°C, after accounting for a 1°C reduction due to room wall orientation differences. These results demonstrate the potential of both systems to enhance thermal comfort, despite increased humidity. Study further simulates thermal comfort achieved by combination façade by iterating the shape factor of clay cones. The combination façade consistently performed better in reducing operative temperatures. Findings validated against ASHRAE Standards highlight the potential of VGS and combination facades, in enhancing indoor environments in warm, humid regions like Trivandrum.

Keyword: *Indoor thermal comfort, Warm humid climate, Vertical Greenery system*

Post-Occupancy Evaluation of Courtyard Spaces in Undergraduate Girls' Hostels: Assessing Occupant Satisfaction

Parul Shelar * , **Dr. A. Meenatchi Sundaram**
National Institute of Technology Tiruchirappalli, Tiruchirappalli, India

Student housing facilities designed with internal courtyards create a comfortable microclimate, significantly influencing student well-being and productivity. This study examines occupant satisfaction with courtyard spaces in undergraduate girls' hostels in the city of Bhopal, evaluating their impact on residents' daily experiences. Using a Post-Occupancy Evaluation (POE) survey, responses from hostel residents were analysed to assess courtyard effectiveness in providing natural lighting, ventilation, seasonal adaptability, spatial efficiency, cleanliness, and safety. Findings indicate 73% of respondents found the courtyard comfortable during winter months, while 49% reported discomfort in summer due to high daytime temperatures. Additionally, 81% of residents recognized the courtyard as a valuable social space, supporting communal activities such as studying, leisure, and cultural events. The study underscores the relevance of courtyards in high-density student housing, particularly in tropical climates, and highlights areas for improvement, including enhanced artificial lighting, accessibility, and maintenance protocols. Insights from this research can inform future hostel designs, ensuring better spatial planning and improved student satisfaction.

References:

Abdul Hadi Nawawi and Natasha Khalil, Post-occupancy evaluation correlated with building occupants' satisfaction: An approach to performance evaluation of government and public buildings, Journal of Building Appraisal VOL.4 NO.2 PP 59–69, (2008)

Schakib-Ekbatan, K and Wagner, A., Assessment of Occupant Satisfaction in Building Performance Evaluation based on Systematic Surveys, (2014)

Yagmur Toprakli, A Post Occupancy Evaluation Study for University Students' Dormitory for User Satisfaction, Journal of Youth Research, (2019)

Keyword: *Post-Occupancy Evaluation, Occupant Satisfaction, Thermal Comfort*

Natural insulation is the way to the future; Analysis of insulation effectiveness, energy consumption, and indoor thermal comfort for future climates

Aritro De * , **Michael Garrison**
The University of Texas at Austin

Natural insulation materials offer promising alternatives to conventional synthetic options for building insulation, providing comparable thermal performance with lower environmental impacts. This study analyzes the effectiveness of various insulation materials, including natural and recycled options, under projected climate conditions, focusing on their impact on building energy performance and thermal comfort. The research employs a multi-stage analytical approach, integrating historical weather patterns with projected climate data to assess insulation effectiveness in a simulated residential building in Austin, Texas. The study examines conventional materials (glass wool, rock wool, expanded/extruded polystyrene), natural materials (flax, hemp, wood fiber), and recycled materials (cellulose, rubber, polystyrene fibers) under current and future climate scenarios (SSP1 and SSP5 for 2050 and 2099). Results indicate that natural and recycled insulation materials can offer comparable and in some cases superior, thermal performance to conventional options under current climate conditions. In future scenarios, all materials show increased cooling demands and decreased heating requirements, with natural materials maintaining their competitiveness. The analysis of Mean Radiant Temperature trends reveals a clear warming pattern, with more extreme daily temperature variations and frequent heatwave events in future scenarios, particularly under SSP5. These findings highlight the potential for sustainable insulation materials to play a significant role in future building designs, offering both environmental benefits and energy efficiency in the face of changing climate conditions.

References:

Calama-González, C.M., León-Rodríguez, Á.L. and Suárez, R. (2023) ‘Climate change mitigation: thermal comfort improvement in Mediterranean social dwellings through dynamic test cells modelling’, International Journal of Energy and Environmental Engineering, 14(2), pp. 121–134. Available at: https://doi.org/10.1007/s40095-022-00498-1.

Belcher, S., Hacker, J. and Powell, D. (2005) ‘Constructing design weather data for future climates’, Building Services Engineering Research and Technology, 26(1), pp. 49–61. Available at: https://doi.org/10.1191/0143624405bt112oa.

Kumar, D. et al. (2020) ‘Comparative analysis of building insulation material properties and performance’, Renewable and Sustainable Energy Reviews, 131, p. 110038. Available at: https://doi.org/10.1016/j.rser.2020.110038.

Keyword: *Natural insulation, Climate change, Energy performance*

An AI-ML based Decision Support Tool for Optimizing Green Infrastructure to Mitigate Air Pollution in Urban Built Environments

Cherith Reddy Yerabolu, Geethika Choudhary Yadlapalli, Monisha Kollipara, Sri Dhruti Tikkisetti, Surekha Tetali *

Mahindra University

One critical impact of climate change, driven by rising greenhouse gas emissions, is its adverse effect on air quality, which directly impacts human health and well-being. While addressing primary sources, such as emissions from energy production and transitioning to electric vehicles, is crucial, decentralized solutions within the urban built environment offer significant potential to mitigate air pollution and enhance resilience. Research has highlighted the effectiveness of nature-based solutions, particularly green infrastructure, in reducing air pollution. However, the efficiency of such interventions often varies based on building designs, systems, and contextual factors. This study addresses this gap by integrating insights from existing research and contextual data to optimize green infrastructure for air pollution reduction in and around buildings. A comprehensive database will categorize building systems and green infrastructure strategies based on their pollution reduction potential. Using text-mining pipelines with Natural Language Processing (NLP) and Machine Learning (ML), the database will synthesize data from peer-reviewed literature and product specifications to systematically catalog building systems and green infrastructure solutions. By integrating these datasets with localized air quality data, a multi-step optimization process will leverage ML and reinforcement learning to identify the optimal combination of building systems and green infrastructure for achieving the lowest possible air quality index. This decision support tool will empower homeowners, builders, architects, urban planners, and policymakers to make data-driven decisions, advancing sustainable urban development and improving indoor and outdoor air quality, fostering healthier and more resilient communities.

Keyword: *Green Infrastructure, Optimization, Air Pollution*

Optimizing energy efficiency and Indoor Air Quality: A Typology-Based Analysis of the European Building Stock

Lisa Corneillie * , **Jelle Laverge**
Ghent University, Ghent, Belgium

In response to updated indoor environmental quality (IEQ) and indoor air quality (IAQ) requirements in the recast Energy Performance of Buildings Directive (EPBD), mapping IAQ across Europe at the building stock level is crucial. This effort is critical for improving energy efficiency, reducing building-related carbon emissions, and enhancing indoor air quality, aligning with the EU's broader climate and health objectives. To analyze the diverse European residential building stock, this study uses the TABULA database, developed in 2012 under the Intelligent Energy Europe (IEE) Project, to define standardized typologies. Buildings are categorized into four groups: single-family homes, terraced houses, multifamily homes, and apartment buildings. Each category is further classified based on common plan compositions and proportions, resulting in 84 representative typologies: 30 single-family homes, 16 terraced houses, 19 multifamily buildings, and 19 apartments. Additionally, an analysis of construction elements—walls, floors, roofs, and windows—led to four insulation scenarios, ranging from poorly insulated to modern standards. These typologies offer a streamlined yet comprehensive model of the European residential stock.

Future research will apply dynamic simulations to evaluate residential ventilation systems, optimizing energy efficiency while improving IAQ. This approach will identify solutions that align with EPBD objectives and EU goals of reducing carbon emissions and promoting healthier indoor environments. Beyond ventilation, this framework serves as a resource for analyzing energy efficiency strategies, optimizing building systems, and supporting evidence-based policy development across the European residential sector.

References:

Energy Performance of Buildings Directive', accessed 20 November 2024, https://energy.ec.europa.eu/topics/energy-efficiency/energy-efficient-buildings/energy-performance-buildings-directive_en.
TABULA_FinalReport.Pdf', accessed 20 November 2024, https://www.episcope.eu/downloads/public/docs/report/TABULA_FinalReport.pdf.

Keyword: *Building stock analysis, residential ventilation*

The Impact of Industrial Waste Dumpsite on Indoor and Outdoor Environmental Air Quality of Ibusu-Asaba and its Environ

Gabriel Friday Ibeh * [1], **Chiazor Stephen Ngozi-Chika**[2], **Oluyemi Toyinbo**[3], **Angelica Ogbimi Emoefe**[2], **Morrison Tamaratimipre Jagboma**[2]

[1] *Department of Physics, Faculty of Science, Dennis Osadebay University Asaba Delta, Nigeria*

[2] *Department of Geology, Faculty of Science, Dennis Osadebay University Asaba Delta, Nigeria*

[3] *Civil Engineering research unit, University of Oulu, Finland*

Industrial waste dumpsites can significantly degrade both indoor and outdoor air quality in their surroundings. Several studies conducted in different locations have revealed similar trends regarding the types of pollutants and their impact on air quality. This study aims to determine the impact of industrial waste dumpsite on indoor and outdoor air quality in the Ibusu-Asaba area.

Dumpsites emit various harmful outdoor pollutants, such as carbon monoxide (CO), hydrogen sulfide (H_2S), methane (CH_4), particulate matter ($PM_{2.5}$), sulfur dioxide (SO_2), and nitrogen dioxide (NO_2). Observations from field investigations have shown that the industrial waste dumpsite in Ibusu-Asaba has the potential to emit these harmful pollutants. These pollutants are released through the decomposition of waste and, in some cases, through fires at the dumpsite.

It has also been observed that many residential houses are clustered around the dumpsite, posing a challenge to air quality, particularly in densely populated residential areas. This proximity increases the risk of exposure to pollutants. Pollutants from this dumpsite can infiltrate indoor environments, especially in residential buildings located nearby. The presence of pollutants indoors, emitted from the waste dumpsite in Ibusu-Asaba, can cause similar health problems as outdoor pollutants. These health issues include respiratory problems and other ailments, particularly affecting vulnerable populations such as young children and the elderly.

Keyword: *Pollutants, Environment, Dumpsite*

Thermal comfort and indoor air quality survey of air conditioned cars in India

Snehangshu Mondal * [1], **Asit kumar Mishra**[2], **Maddali Ramgopal**[1], **Siddhartha Mukhopadhyay**[1]
[1] *Indian Institute of Technology Kharagpur*
[2] *University College Cork, Ireland*

Currently almost all the passenger cars produced in India are provided with Heating Ventilation and Air Conditioning (HVAC) systems for thermal comfort. In addition to thermal comfort, the HVAC systems are expected to provide proper Indoor Air Quality (IAQ) also. However, these energy intensive car HVAC systems affect the fuel consumption and/or the range of the cars significantly, in addition to contributing to the global warming and other environmental issues. Hence, it is essential to ensure that the additional energy is well-spent in an environment friendly manner, apart from meeting the intended purpose of providing thermal comfort and air quality to the passengers. This requires detailed theoretical and experimental studies and subjective surveys. However, such studies are very scarce, especially for Indian cars and Indian conditions. Hence to assess the thermal comfort and indoor quality aspects of passenger cars, a subjective survey was carried out through a well-designed questionnaire. The questionnaire captures, apart from basic occupant details such as the gender, age and seating position, the occupants' response to temperature, humidity, air motion and air quality. A total of 235 responses were received, covering a variety of occupants and passenger cars. The analysis of the survey results highlights interesting aspects about the passenger car air conditioning systems and also points out directions for improving the energy performance of the car HVAC systems without affecting the passenger comfort.

References:

Corgnati, S. P., Filippi, M., & Viazzo, S. (2007). Perception of the thermal environment in high school and university classrooms: Subjective preferences and thermal comfort. Building and environment, 42(2), 951-959.

Keyword: *Thermal comfort, Air quality, Air conditioned cars, Questionnaire, Field survey*

Thermal comfort and indoor air quality survey of air conditioned cars in India

Snehangshu Mondal[1*], Asit Kumar Mishra[2], Maddali Ramgopal[3], Siddhartha Mukhopadhyay[4]

1 School of Energy Science and Engineering, Indian Institute of Technology Kharagpur-721302, India

2 School of Public Health, University College Cork, Ireland

3 Department of Mechanical Engineering, Indian Institute of Technology Kharagpur-721302, India

4 Department of Electrical Engineering, Indian Institute of Technology Kharagpur-721302, India

(*Corresponding Author: snehangshumondal.biet93@kgpian.iitkgp.ac.in)

SUMMARY

A subjective survey was carried out to assess the thermal comfort perception of passengers of air conditioned cars in India. The survey questionnaire prepared captures data on the gender and age of the passenger, his/her seating position, perception about cabin air temperature, moisture content, air velocity, indoor air quality etc. The thermal comfort survey questionnaire is framed in Google forms and is shared through email and social apps across the country. A total number of 235 responses were collected within the given survey period. Survey results show that out of the total occupants, more than 70% expressed satisfaction with the air conditioning system, even though they aspire for better systems. during the months of summer season in India. An analysis of the survey results show that around 60 % of the passengers feel that they will be comfortable in the temperature range of 22°C to 26°C. Though majority of the passengers seem to be comfortable with humidity and air motion inside the car cabin, there is some dissatisfaction in terms of stuffiness and sweatiness inside the car cabin, which need to be addressed through better air distribution systems. No major difference in terms of age group or gender could be observed from the survey results.

Keywords: Thermal comfort; Air conditioned cars; Passenger perception; Subjective survey

1 INTRODUCTION

The exponentially growing needs of efficient air conditioning (AC) system in cars and buildings can be ascribed to worsening climatic conditions, growth of population, urbanization and industrialization in India. With rising standards of living, there is a notable increase of vehicles on the Indian roads in the recent decades. Most of the passenger cars that are sold now are equipped with AC systems. While, the need for air-conditioned cars for passenger thermal comfort cannot be questioned in view of the India's heterogeneous climatic conditions, the associated fuel costs and greenhouse gas emissions are of major concern. Studies show that a hot and humid environment can negatively impact occupant comfort and impair driver awareness and cognitive performance [1], outlining the need for an efficient air conditioning systems. The rise of electric vehicles (EVs) with constraints on battery capacity poses new challenges in the design of air conditioning systems [2]. A survey in India during peak summer and rainy seasons revealed significant discrepancies between the manufacturer-rated AC performance, range, and mileage versus actual on-road data. User-reported statistics from automotive forums highlight concerns about slow temperature stabilization, high energy use in traffic, and poor rear-seat cooling [3]. Modern vehicular climate control systems must balance fast cabin cooling, energy efficiency for EV range, and air quality management in urban areas with high PM2.5 levels [4]. Traditional thermal comfort assessment methods, suited for stationary buildings are inadequate for automotive applications due to confined cabin space, dynamic solar radiation, and varying occupant metabolic rates [5]. Studies show that the Predicted Mean Vote (PMV) index needs adjustment for automotive environments due to microclimate changes from localized air velocities and significant radiant asymmetry between sun-exposed and shaded surfaces [6]. The thermal environment inside automobile cabins can be assessed through suitable computational fluid dynamics (CFD) models. However, subjective surveys are needed for assessing the thermal comfort perception of the occupants of a stationary building or a moving car. The purpose behind thermal comfort surveys in vehicles and buildings is the same, i.e., to gauge occupant satisfaction with their thermal environment, but the procedures and impacting variables could be different. Branco et al., [7] monitored indoor air quality in real-time during 40 trips, selecting 24 to study factors like pollution, traffic intensity, ventilation strategies, and parking garage circulation. Route specifics, weather, and traffic intensity were also logged for detailed analysis. Ye et al., carried out field surveys in a long distance sleeper train in China to assess the indoor environment and air quality [8]. The study combines objective measurements and subjective surveys to assess the impact of thermal zone transitions on occupant temperature perception. Research on thermal comfort in passenger rail trains highlights the importance of personal comfort zones and individual climate control. Studies, including the one with 172 participants, emphasize the role of "Thermo Specific Self Efficacy" (TSSE) in improving comfort, and recommend integrating TSSE into comfort forecasts [9].

Research Gaps

Though thermal comfort surveys have been conducted in various types of buildings in India, similar studies for mobile applications such as passenger cars are not reported in open literature for Indian conditions.

Purpose of the Survey

The primary aim of the survey is to get an overall idea about the perception of the passengers on thermal comfort aspects of cars running on Indian roads. The survey is designed to collect the subjective perceptions of the occupants regarding the air temperature, humidity level, air velocity, indoor air quality, and to see whether there is any relation between their seating position, gender and age on these perceptions. It is a preliminary study that will be followed more detailed subjective surveys, numerical and experimental studies with a goal to come up with recommendations for efficient design and operation of passenger air conditioning systems.

2 METHODOLOGY

The thermal comfort survey of air conditioned cars is done by preparing a questionnaire that captures information such as type of passenger car, gender and age of the passengers, their seating position, their thermal comfort perception, preferred temperature range, perceived air motion, humidity level, indoor air quality and journey details.

As the survey is carried out by circulating the questionnaire prepared using Google Forms through emails and social apps, no environmental parameter mapping is done. It is assumed that the response to the questionnaire by the respondents is an average response based on several of their trips in air conditioned cars. Thus the conclusions drawn out of the survey results provide an overall perception of the passengers on air conditioned cars, and not meant for generating any explicit correlations between the comfort parameters and the associated perception.

Location, Types of vehicle and Subjects

Location: The thermal comfort survey of the AC cars I conducted in India.

Types of vehicle: Five types of AC cars are considered in this study. These are: Micro, Hatchback, Sedan, SUV and Wagon as shown in Fig. 1.

Subjects: The respondents are a mix of female, male and other genders. The age groups considered are from 19 years to 65 years.

Fig. 1: Various types of air conditioned cars

Data collection: During the survey no physical parameters were measured. Only Google form based subjective questionnaires were filled by the occupants hailing from different regions of the country who experienced travel in AC cars.

Subjective questionnaire: The questionnaire begins with a brief introduction and purpose behind the subjective survey, followed by questions related to various aspects of the passengers and their perception about the environment inside the car cabin. The survey questionnaire was made in three different languages, namely, English, Hindi and Bengali. As mentioned, the survey questionnaire was prepared in Google Forms and circulated through email and social apps. The link for the questionnaire is given below:

Link for the Google Forms

The respondents filled the questionnaire based on their present and past travel experience in air conditioned cars and similar vehicles on trips that last 30 minutes or longer.

The responses were collected and exported to MS Excel sheet and RStudio for analysis.

3 RESULTS AND DISCUSSION

From the responses received it is observed that approximately 47% respondents are from the state of West Bengal, which typically has a hot and humid climate. Out of the 235 responses, around 14% (33) are female respondents and the rest are male respondents. In terms of age group, around 17% are in the age group of 19-25 years, around 51% respondents fall in the age group of 26-35 years, 17% are in the age group of 36-45 years and the remaining (about 15 %) are above 45 years. Out of the total responses, 50% mentioned hatchback type of cars and 36% mentioned sedan type of cars. Out of the 235 respondents, around 69% occupants travel by sitting in either driver's seat (around 24%) or in the front passenger seat (around 45%). Approximately 48% women travel by either hatchback car or sedan car, whereas 47% and 32% men travel by hatchback car and sedan car, respectively. Most of the low price range sedan and hatchback cars have air distribution port at the dashboard.

Out of the total occupants, around 75% expressed that they are thermally comfortable inside the air conditioned cars, while around 17% felt neutral. Around 8% of the occupants, however, mentioned that they are not comfortable with the temperature inside the AC car. The percentage of male passengers who expressed dissatisfaction with the inside temperature is approximately same as that of the female passengers. In terms of the seating position, 84% occupants in the driver's seat expressed satisfaction with thermal comfort conditions, whereas around 72% sitting in passenger seats (front and back) are satisfied with the thermal comfort provided by the car air conditioning system.

In terms of preferred temperature range, 35% respondents preferred the 22°C - 24°C range, while around 26% respondents are thermally comfortable within the 24 °C - 26°C band. Thus the preferred temperature range for over 60 % of the occupants, irrespective of their gender or seating position is within 22°C - 26°C, as shown in Fig. 2. Around 12 % of the passengers preferred a much lower temperature range of 18°C-20°C. Only around 7 % of the occupants preferred temperatures above 26°C. In terms of the preferred temperature range, no major difference could be observed between male and female passengers or among the different age groups.

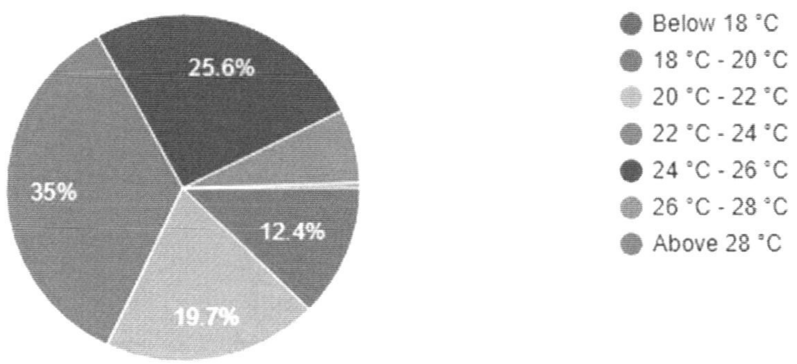

Fig. 2: Percentage of occupants with preferred temperature range

Survey results show that changing the direction and magnitude of air velocity with the help of the louvres, change of blower speed, and change of temperature set points are the most commonly used methods by the passengers to achieve thermal comfort.

Results show that in 91.5% of the cases there are no extra fans mounted on the frames or panel inside the car. However, around 61% of the occupants are satisfied with the air movement inside the vehicle.

Fig. 3 shows the perception of the passengers to humidity levels inside the car. It can be seen that majority of the passengers appear to be satisfied with the humidity level inside the cabins, with less than 20% reporting very humid, humid, dry or very dry conditions. Again there is no clear distinction between male or female passengers or between front or back seat passengers in terms of their perception about humidity.

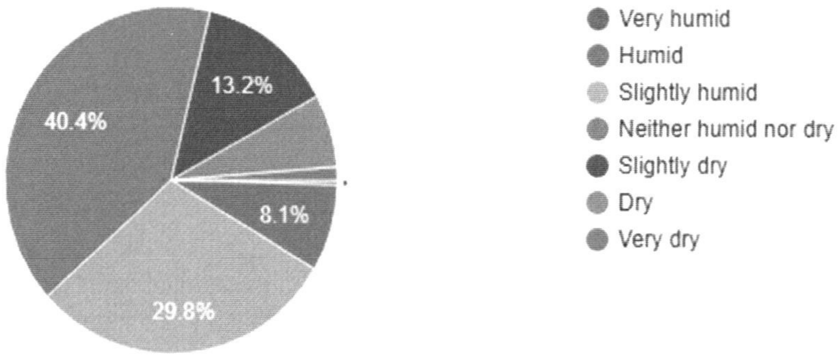

Fig. 3: Percentage of occupants with their perception about humidity

Out of the total occupants, 42% are neutral to the air quality inside the car, but around 36% occupants reported slightly stuffy conditions while around 4% recorded very stuffy conditions. While there is no difference between male and female passengers or different age groups, in terms of their perception about stuffiness inside the car, compared to the front seat passengers, higher number of passengers sitting in the rear seat felt more stuffy.

In terms of sweatiness at the back and glute, around 18% of the passengers reported sweatiness. Out of the passengers reporting sweatiness, around 54% are male, while the remaining or female.

In terms of overall satisfaction, around 72% occupants voted for overall satisfaction but seek better comfort. Around 11% occupants are not satisfied and they asked for improvement of the car AC and air distribution system. Some occupants suggested AC vents at the rear side, with higher airflow at the rear seats, automatic control as per the outside temperature and an extra fan inside the cabin.

A correlation matrix is generated among the independent variable which effect the thermal comfort by generating heat map using R studio, shown in Fig. 4. The following observations are found based on the heat map: High Positive Correlation (Darker Shades to Deep Red): Humidity Level ↔ Air Inside the Car: Because moist air feels heavier and less fresh, higher humidity is substantially associated with a worse perception of air quality.

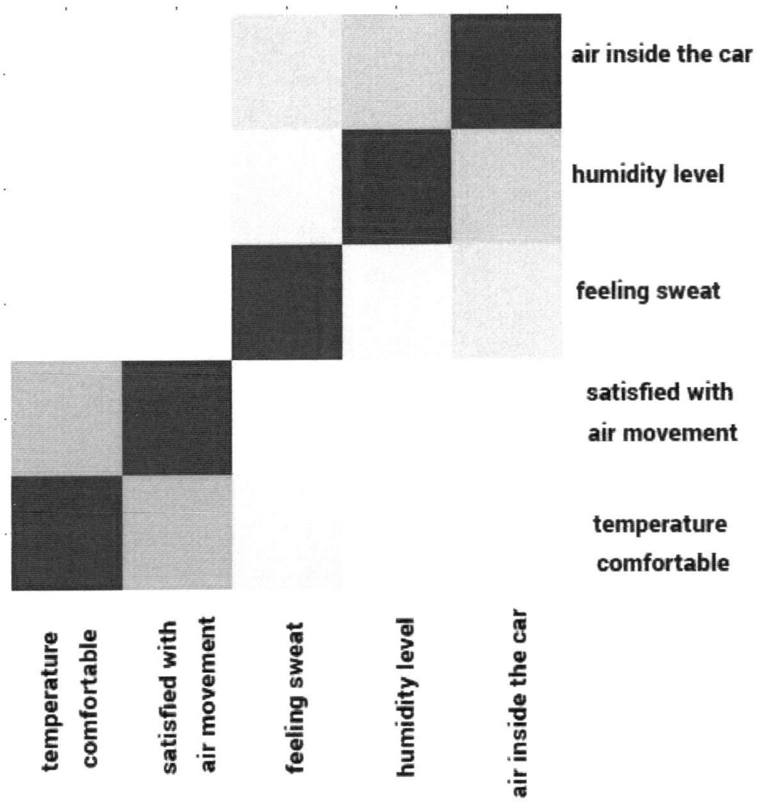

Fig. 4: A heat map correlation matrix for many aspects of thermal comfort

Moderate Positive Correlation (Medium Shades: Orange to Red): Sweating ↔ Humidity Level: Higher humidity, as predicted, tends to cause more perspiration, which lowers comfort levels. Feeling Sweat: Low or No Correlation (Lighter Shades-Yellow): Comfortable Temperature Air circulation, individual tolerance, and cooling effectiveness are likely to have an impact on thermal comfort, which is not always directly related to perspiration. Principal Finding: Pay attention to airflow and humidity regulation: Since ventilation and humidity have a big impact on how uncomfortable something feels, they can greatly increase overall thermal comfort.

Preferred Temperature Range and Air Movement Satisfaction are more consistently rated, reflecting clearer passenger preferences, while Humidity Level and Sweat Feeling have the most varied responses, suggesting they are more context-dependent, according to the Distribution of Key Thermal Comfort Factors, shown in Fig. 5. The disparity in thermal comfort and overall experience emphasizes how crucial efficient climate management is to passenger comfort.

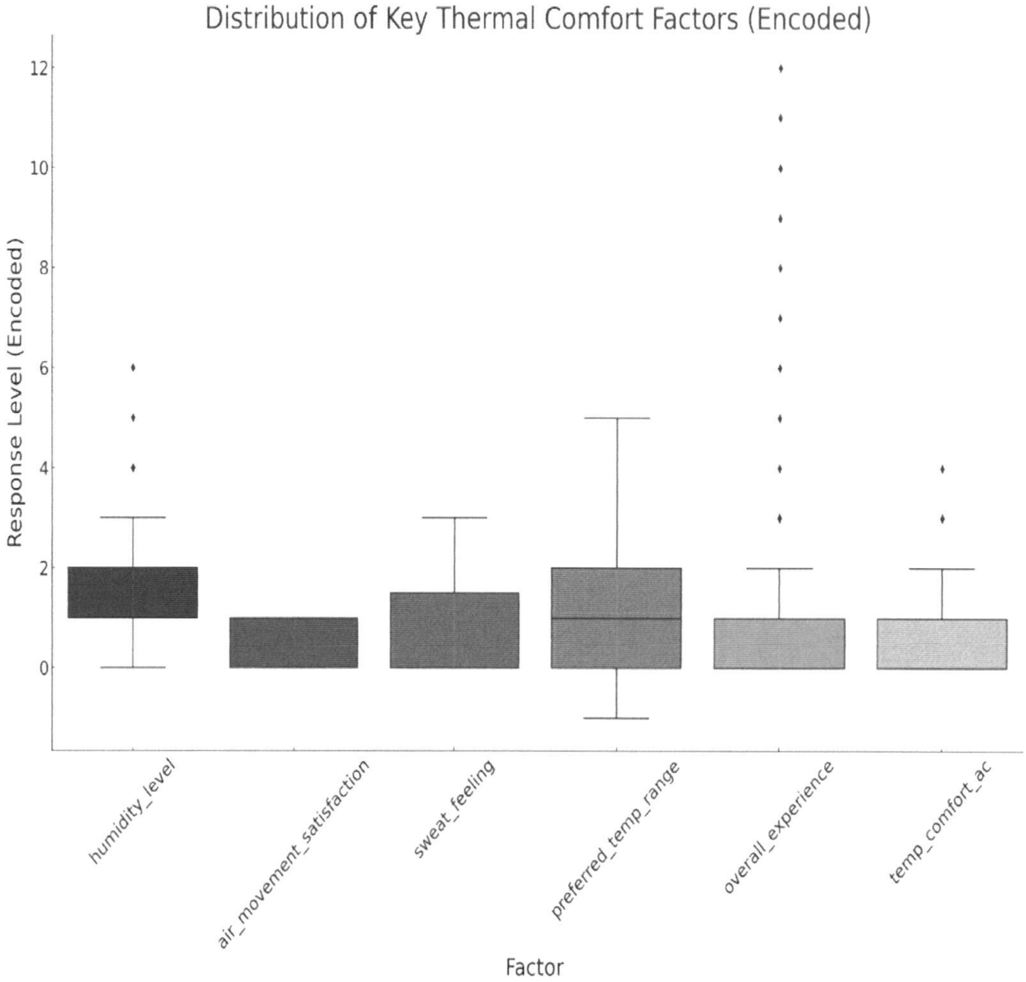

Fig. 5: Distribution of Key Thermal Comfort Factors

4 CONCLUSIONS

The preliminary survey carried out on the thermal comfort perception of passengers travelling in cars on Indian roads showed some interesting trends. It is seen that over 70 % of the passengers who participated in the survey expressed satisfaction with the air conditioning system in the cars, even though they aspire for more comfort. The preferred temperature range for majority of the passengers is in the 22°C to 26°C range. No major difference in perception is observed in terms of the age groups or gender. In terms of seating position, a higher % of occupants in the driver position expressed satisfaction compared to others. This could be due to the greater control the driver exercises for changing the temperature and air velocity settings. A heat map correlation matrix is generated among the independent variables to identify the effect on thermal comfort perception. Preferred temperature and air movement are the key factors for thermal comfort perception based on the thermal comfort analysis and these are consistently rated based on the box plot. There are some concerns about the stuffiness and sweatiness inside the car cabin, which may have to be addressed through better air distribution inside the cars.

5 REFERENCES

[1] Ravindra, K., Agarwal, N., & Mor, S. (2020). Assessment of thermal comfort parameters in various car models and mitigation strategies for extreme heat-health risks in the tropical climate. *Journal of Environmental Management*, *267*, 110655.

[2] Basciotti, D., Dvorak, D., & Gellai, I. (2020). A novel methodology for evaluating the impact of energy efficiency measures on the cabin thermal comfort of electric vehicles. *Energies*, *13*(15), 3872.

[3] Zhang, L., Qi, L., Liu, J., & Wu, Q. (2020). Experimental study on dynamic thermal environment of passenger compartment based on thermal evaluation indexes. *Science Progress*, *103*(3), 0036850420942991.

[4] Wei, D., Nielsen, F., Karlsson, H., Ekberg, L., & Dalenbäck, J. O. (2023). Vehicle cabin air quality: influence of air recirculation on energy use, particles, and CO2. *Environmental Science and Pollution Research*, *30*(15), 43387-43402.

[5] Luo, J., Chen, G., Xie, P., Zhang, H., Xu, S., Chen, X., ... & Jiang, C. (2024). Research progress of human thermal comfort in vehicle passenger compartment based on computational fluid dynamics. *Proceedings of the Institution of Mechanical Engineers, Part D: Journal of Automobile Engineering*, 09544070241306826.

[6] Jose, S. S., & Chidambaram, R. K. (2021). Thermal Comfort Optimization in an Electric Vehicle. *International Journal of Heat & Technology*, *39*(6).

[7] Branco, P. T., Pinto, A. J., & Sousa, S. I. (2024). Characterising indoor air quality in private vehicle cabins under unprecedented traffic conditions during COVID-19 lockdown. Building and Environment, 262, 111797.

[8] Ye, X., Lu, H., Li, D., Sun, B., & Liu, Y. (2004). Thermal comfort and air quality in passenger rail cars. International Journal of Ventilation, 3(2), 183-192.

[9] Maier, J., Zierke, O., Hoermann, H. J., & Goerke, P. (2021). Effects of personal control for thermal comfort in long-distance trains. Energy and Buildings, 247, 111125.

Interpretable AI Model for Indoor Thermal Comfort Prediction: A SHAP-Based Analysis Using Benchmark Dataset

Simon Elias * [1], **Iyer Vijayalaxmi Kasinath**[2]
[1] *School of Planning and Architecture, Vijayawada*
[2] *Professor, Dept. of Architecture, School of Planning and Architecture, Vijayawada.*

Accurate prediction of indoor thermal comfort is crucial for optimizing building energy consumption and enhancing occupant well-being [2,3]. AI models are increasingly employed to predict thermal comfort, supported by benchmark datasets for standardization. The interpretability of predicted values, including the factors and their significance, provides deeper insights into optimizing building energy use [1].

This study introduces an approach for deriving interpretations from AI models trained to predict the Predicted Mean Vote (PMV), a standard index for evaluating thermal comfort. The ASHRAE II dataset was used to train an XGBRegressor (Extreme Gradient Boosting Regressor) machine learning model. To enhance interpretability, SHAP (SHapley Additive exPlanations) was applied to the trained model to elucidate feature contributions. Interpretable AI facilitates the identification of key drivers of thermal comfort PMV predictions.

An XGBRegressor model was trained on the ASHRAE II dataset using 14 key features known to influence PMV predictions. The model achieved a R^2 score of 0.9555, explaining 95.55% of the variance in PMV, with a Root Mean Square Error (RMSE) of 0.1567 PMV, indicating an average deviation of ±0.1567 units from actual values. Given that PMV ranges from -3 (cold) to +3 (hot), this low error suggests the model's strong accuracy in thermal comfort prediction.

SHAP analysis revealed a positive relationship between feature values and their contributions to the model's predictions. Furthermore, the analysis showed that lower feature values resulted in minimal or negligible contributions to the predicted outcome. These findings highlight the importance of feature scaling in thermal comfort modeling.

References:

Yuanchuan Li, Feng Gao, Jiayue Yu, Teng Fei, Machine learning based thermal comfort prediction in office spaces: Integrating SMOTE and SHAP methods, Energy and Buildings,Volume 329, 2025, 115267, ISSN 0378-7788.
Nourhan M. Waly, Hamdy Hassan, Ryo Murata, David J. Sailor, Hatem Mahmoud, Experimental approach to reduce uncertainty in estimating the thermal performance of buildings' envelopes in hot climates, Developments in the Built Environment, Volume 21, 2025, 100589, ISSN 2666-1659.

Keovathana Run, Franck Cévaër, Jean-François Dubé, Influence of transient conditions on thermal comfort evaluations: Field study in university buildings, Case Studies in Thermal Engineering, Volume 66, 2025, 105737, ISSN 2214-157X.

Keyword: *Thermal Comfort, Interpretable AI, ASHRAE*

CFD Validation of HVAC System Design with Fabric Ducts for Bio Domes at Doha Expo-2023: Achieving Optimal Indoor Environmental Conditions

Hashir PK * [1], **Veerasingam Subramanian**[2], **Fadil Sadooni**[2], **Saud Ghani**[1]
[1] *Mechanical and Industrial Engineering, Qatar University*
[2] *Environment Science Center, Qatar University*

This study conducts a steady state Computational Fluid Dynamics (CFD) analysis to validate the performance of an HVAC system for Bio Domes, addressing the need for proper air distribution within the indooor space. A steady-state, three-dimensional RANS approach with the realizable k-epsilon turbulence model was used to simulate turbulent airflow using ANSYS. The energy equation would be solved to account for heat transfer, and a multi-species transport model was used to determine the relative humidity. The study employs a detailed 3D model of the dome structure of a hemispherical shape with 30 m diameter and 15 m height, incorporating equipment specifications and heat load assessments, to simulate fluid flow and heat transfer under various conditions. Consistent with recent research, the analysis evaluates the impact of system components on indoor thermal parameters in controlled environments. Three air handling units (AHU) make up the HVAC system. Their combined airflow of 38,250 CFM is dispersed by a combination of 2-inch and 4-inch nozzles in the fabric ducts. Important conclusions drawn from the simulation show that the HVAC system successfully keeps the relative humidity at 51.8% and the average indoor temperature at 21.1°C. These findings show low vertical fluctuations in temperature and humidity levels throughout the dome, guaranteeing even air distribution and occupant comfort. The study highlights the effectiveness of fabric duct systems in large-volume structures like Bio Domes and underscores the importance of CFD in optimizing energy efficiency and environmental control in modern building systems. The integration of CFD analysis with experimental data further enhances result reliability by validating simulations through measurements of temperature differentials, airflow rates, and power consumption. This combined approach provides valuable insights into thermal performance and supports the development of more efficient HVAC solutions for large-scale applications.

Keyword: *Computational Fluid Dynamics (CFD), Indoor Thermal Comfort, Fabric Duct*

Assessing the energy saving potential of personalized ventilation system in the Indian tropical climate

Kumar Naddunuri * [1], **Shankha Pratim Bhattacharya**[2]

[1] *Department of Architecutre and Regional Planning, Indian Institute of Technology, Kharagpur, India*

[2] *Department of Architecutre and Regional Planning, Indian Institute of Technology, Kharagpur*

This study aims to evaluate personalized ventilation system energy-saving potential in the Indian context, India's climate is characterized by high temperatures and humidity levels making personalized ventilation an attractive option for energy-efficient cooling. Personalized ventilation (PV) system also known as a personal environmental control system (PECS) targets to cool the occupant micro-region instead of the entire space, the system effectively delivers fresh air close to the occupant's breathing zone, resulting in higher ventilation efficiency and acceptance of thermal environment due to the system's ability to provide controls over supply conditions. Studies have shown that operating a personalized ventilation system when coupled with different background ventilation systems can yield better energy savings, In Indian tropical climates, understanding the occupant's preferred comfort bands, air movement, behavioural patterns, and background conditions and adapting to use the PV system efficiently in practice can greatly impact the overall energy consumption of air conditioning system. This paper employs an analytical method to discuss the acceptable thermal environment under personalized ventilation, occupant behavioural patterns, and their background conditions to assess the personalized ventilation system energy saving potential. The study underlines the previous findings that occupants are tolerant at slightly high background temperatures, however, the behavioural aspects of the occupants contribute to high energy consumption due to their ability to adapt to the system quickly. Therefore, modifying or adjusting certain behavioural styles while using the personalized ventilation system controls effectively can lead to energy benefits through informed decisions.

References:

Melikov, A. K. (2004). Personalized ventilation. Indoor air, 14.

Kaczmarczyk, J., Melikov, A., & Fanger, P. O. (2004). Human response to personalized ventilation and mixing ventilation. Indoor Air, 14.

Kaczmarczyk, J., Melikov, A., Bolashikov, Z., Nikolaev, L., & Fanger, P. O. (2006). Human response to five designs of personalized ventilation. Hvac&R Research, 12(2), 367-384.

Keyword: *Energy efficient cooling, Thermal environment, Occupant behaviour, Ventilation efficiency, Energy consumption*

Role of Sacred groves in maintaining a sustainable ecosystem in Traditional Homesteads of Kerala-A case of Trivandrum

Radhika Radhakrishnan * , **Dr. Deepa Rani R**
College of Engineering Trivandrum

Kerala, a state in India renowned for its lush greenery, a unique environmental conservation practice – sacred groves, locally called kaavu. These patches of forests are protected not by government regulations but by deep-rooted cultural and religious beliefs. These groves are integral to the traditional homesteads of Kerala, playing a crucial role in maintaining the ecological balance and supporting biodiversity. This study helps in understanding the influence of sacred groves and the adjoining ecosystem on outdoor thermal comfort in traditional homesteads of Kerala also safeguarding Kerala's rich biodiversity.

Kaavu represents a unique blend of cultural heritage and environmental conservation, embodying principles of sustainability that have been practiced long before modern environmental movements. These dense pockets of vegetation act as carbon sinks, absorbing and storing atmospheric carbon dioxide, a key greenhouse gas. Additionally, kaavu promotes soil conservation, prevents erosion, and regulates local water cycles, all of which contribute to a more resilient ecosystem in the face of a changing climate.

This study continues with onsite measurements obtained from morning 9am, 11am, 1pm and 4pm around the areas within sacred groves and an extended larger context in order to understand the impact. Outdoor thermal comfort parameters such as temperature, humidity, mean radiant temperature, and air velocity were measured and validated compared to simulations. By analyzing the role of kaavu in maintaining a sustainable ecosystem in Traditional Homesteads of Kerala, this study positions them as a powerful model for microclimate modification.

References:

Chandrasekhara, U.M2011. Conservation and Management of Sacred groves in Kerala. (KFRI Research Report No: 412: p)
Jeevan, Kavya Bonny, Binoo PS, Gopakumar. (2023) Climate change perspectives and conservation of sacred groves: Case of Sharngakavu, Kerala
U.M. Chandrashekara S. Sankar (1998). Ecological and Social Importance of Conservation of Sacred Grooves In Kerala

Keyword: *Sacred groves, Homesteads, Carbon sinks*

Thermal comfort of adolescent children – Engaging children for improving reliability of responses

Asit Kumar Mishra * [1], **Pawel Wargocki**[2], **Alice Caporale**[3]
[1] *University College Cork*
[2] *Technical University of Denmark*
[3] *University of Bologna*

Classroom thermal environment is not only important for comfort of students, it is also impacts the teaching-learning process. Existing research on thermal comfort in classrooms for adolescent children shows that their thermal comfort needs are different from adults. This underlines the need for a greater focus on thermal comfort research involving adolescents. However, such works cannot directly extrapolate questionnaires used for adults to the children. Such an extension of these study instruments is liable to confuse children and their responses.

In this work, we describe the methodology we adopted in a living labs study, involving 8-11 years old school children, to engage them and gain more reliable responses. With a total of 20 participants, the engagement strategies included descriptive videos related to indoor environment and thermal comfort, questionnaires that used pictorial representations for the scale points, and treating children as important participants in the process. This led to children remaining engaged through the entire study duration of close to four hours and being willing respondents to the thermal comfort related questionnaires. Cronbach's alpha for the cluster in the questionnaire related to wellbeing was evaluated at 0.87 while the value for the cluster related to thermal comfort-acceptability was 0.89, indicating high internal consistency for the children's responses. The engagement instruments were created in both English and Danish. The protocol generated through this work can be extended to other similar research involving young children. Future works would need to tailor these engagement instruments for different age groups.

Keyword: *thermal comfort, engagement, thermal comfort questionnaire*

Development of Adaptive Thermal Comfort Model and Comfort Band for Office Buildings in Kathmandu, Nepal

Prativa Lamsal * [1], **Sushil Bahadur Bajracharya**[2], **Hom Bahadur Rijal**[3], **Richard de Dear**[4]

[1] *Department of Architecture, Pulchowk Campus, Institute of Engineering, Tribhuvan University, Nepal and Sydney School of Architecture, Design and Planning, The University of Sydney, Sydney, Australia*

[2] *Department of Architecture, Pulchowk Campus, Institute of Engineering, Tribhuvan University, Nepal*

[3] *Department of Restoration Ecology and Built Environment, Tokyo City University, Yokohama 224-8551, Japan*

[4] *Sydney School of Architecture, Design and Planning, The University of Sydney, Sydney, Australia*

Thermal comfort requirements in buildings are essential for energy efficiency and productivity since people spend 87% of their time indoors. Nepal's National Building Code does not include thermal comfort standards, developing an adaptive thermal comfort model with clearly defined comfort bands is essential for establishing such standards. So, this study aims to develop an adaptive model and comfort band for office buildings in Kathmandu, Nepal. A monthly transverse survey was conducted from May 2023 to August 2024 from 8 office buildings, collected 2653 thermal comfort responses from free running(FR) and Air-conditioned(AC) modes. Firstly, the probit analysis was used to establish the comfort bands for 80% and 90% considering indoor operative temperature as a covariate. For FR mode, the 80% comfort band is ±5K, while the 90% comfort band is ±3.5K. This wider range is year-round comfort band due to seasonal variations. To identify the comfort band for a specific moment Δt (T_{op} - T_c) was considered as a covariate. Then, the comfort bands for 80% and 90% acceptable limits are ±2.4K and ±2.1K, respectively, for FR mode and ±2.2K and ±1.7K, respectively for AC mode. The adaptive model was also proposed for both free running and air conditioning modes. The regression coefficient 0.35 and 0.16 were obtained for free running and air conditioning modes respectively. This adaptive model can serve as a tool for estimating comfort temperature and a guiding tool for energy-efficient building design in Kathmandu Nepal.

Keyword: *Office building, Comfort band, Adaptive model*

A Personalized PMV-Based Strategy for Coordinating Ceiling Fans and Air Conditioners

Ramandeep Singh, Althaf Shaik, Anupama Kowli *
Indian Institute of Technology Bombay

Heating, ventilation, and air conditioning (HVAC) accounts for approximately 38% of total electricity consumption in buildings. Although a significant fraction of the building demand, HVAC operations may not result in thermal comfort for the occupants. The presence of ceiling fans with air conditioning (AC) systems – a condition common in India – further exacerbates this problem due to lack of coordination between the two systems. This study presents a feedback-based control system that dynamically adjusts AC and ceiling fan settings based on the Predicted Mean Vote (PMV) metric. Temperature and humidity sensors are deployed to facilitate PMV computation while the air velocity required for the same is approximated using results of computational fluid dynamics (CFD) simulations for different conditions. The sensing, computation and actuation is facilitate via the Home Assistant platform.

Additionally, user interactions with the devices are studied to infer user feedback and the desired PMV values indicating preferred/comfortable states for the user are deduced. In this way, the actuation can be configured to drive the real time PMV to the desired state, ensuring a more personalized thermal comfort strategy. This approach requires no additional infrastructure modifications to existing cooling devices, making it a cost-effective and scalable solution for residential applications. The complete system is demonstrated in a lab setting and the energy saving potential of the strategy is investigated.

Keyword: *Air conditioning, ceiling fan, personalized thermal comfort*

Consumer Demand Perception of Thermal Comfort

Shivani Senthilkumar * , **Poulomee Ghosh**
NICMAR University

The increase in urban temperature and heat waves significantly impacts indoor thermal comfort. Indian Cities like Pune, despite being in the moderate climatic zone, are also dealing with the same issue. This study examines the residents' perception of thermal comfort in Pune homes and aims to establish the relationship between the thermal comfort levels and the building elements and demographic factors. For this, surveys were conducted among the multi-story apartment occupants of Pune, and the data were analysed using the repeated-measure ANOVA test in the SPSS tool. The results revealed that more windows and the single glazing type have a significant effect via the ANOVA test. The compared means show that demographic factors like women, homemakers, and middle-income people and building elements like vegetation and ventilation type have higher thermal comfort levels. Thermal Comfort perceived by the occupants is different, even though it gets affected by the age, gender and environmental parameters.

Keyword: *Thermal Comfort, Occupants Perception, Effects of Building Elements*

Investigating Indoor Airflow: Real-Scale Flow Field Measurement Using Streak Velocimetry

Girish Naik * [1], **Sairam Jinnawar**[2], **Manaswita Bose**[1], **Krishnendu Sinha**[2]
[1] *DESE, IIT Bombay*
[2] *AE, IIT Bombay*

Indoor thermal comfort is influenced by key environmental parameters such as temperature, humidity, and air velocity. Proper management of these factors is essential for ensuring occupant well-being and energy efficiency. Air conditioning (AC) systems are commonly used to regulate indoor temperatures, but they often create uneven cooling and localized stagnant air zones. Ceiling fans, by contrast, improve thermal comfort through air circulation, which provides a well-mixed airflow. When used together, AC and ceiling fans can provide uniform temperature distribution, better air mixing, and reduced energy consumption. However, understanding the airflow interactions between these systems and the flow field characteristics remains a challenge.

This study investigates the airflow characteristics of AC, ceiling fans, and their combined operation by analysing flow fields through streak velocimetry (SV). The experimental setup captures air velocity flow fields with individual and combined configurations of AC and ceiling fans. The velocity measurement from SV technique can be used in conjunction with temperature and humidity values to assess human comfort under different operating conditions. This will help in optimizing ventilation strategies to achieve adequate comfort level at minimum energy consumption, with a potential to apply to a wide variety of ventilation scenarios.

Keyword: *Thermal comfort, indoor airflow, streak velocimetry*

DYNAMIC ORIGAMI FACADES: A COMPARATIVE STUDY ON ENHANCING ENERGY EFFICIENCY AND DAYLIGHT OPTIMISATION IN BUILDING DESIGN

Arshad ul Islam N * , **Bimal P**
National Institute of Technology Calicut

The study explores the potential of origami-inspired patterns in enhancing energy efficiency and optimizing daylight intervention in building facades. The primary research question investigates whether dynamic origami patterns outperform conventional secondary walls, such as brise soleil and double-skin facades, in terms of energy efficiency and interior daylight distribution.

The study compares various origami patterns with traditional facade systems through prototypes at different building scales. Prototypes are analysed using computational models to assess thermal performance, energy conservation and daylight penetration. Key parameters, including the kinetic behaviour and fractal geometry of the origami patterns are incorporated into a generative algorithm designed to optimize these patterns for both energy efficiency and occupant comfort.

Preliminary findings suggest that dynamic origami facades provide enhanced energy savings and more effective daylighting compared to traditional double-skin systems, particularly in regions with varying sunlight exposure. The results highlight the potential of origami-inspired patterns to offer a more adaptable and sustainable solution for building facades, demonstrating their ability to improve interior environmental quality while reducing energy consumption.

This research contributes to the field of sustainable architecture by providing a comparative analysis of origami-based dynamic facades and conventional facade systems, offering insights into their practical applications for optimizing building energy performance.

Keyword: *Origami, Dynamic facades, Building performance*

Thermal Comfort Modeling in Split Air-Conditioned Buildings: A Machine Learning Approach

Ajith N Nair * [1], **Prashant Anand**[1], **Ardeshir Mahdavi**[2]
[1] *Indian Institute of Technology, Kharagpur*
[2] *Institute of Building Physics, Services, and Construction, Graz University of Technology, Austria*

Thermal comfort assessments are crucial for occupant satisfaction and optimizing building energy consumption. Thermal comfort in buildings is commonly evaluated using the PMV model for air-conditioned buildings and the Adaptive Comfort model for naturally ventilated buildings, both of which were developed based on laboratory experiments and field data. However, these models often fail to predict thermal comfort precisely in real-world conditions. This may be in part due to contextual factors, occupant behavior, and other dynamic parameters. In this context, Machine Learning (ML) can aid thermal comfort predictions, leveraging AI technologies to model the complex nonlinear relationships within empirical data. This study employs ML algorithms to predict Thermal Comfort Votes (TCV) and Thermal Acceptance Votes (TAV) using experimental data collected from a split air-conditioned educational building in India. Input for the models includes environmental, personal, and air quality parameters. Five ML models were evaluated: Gradient Boosting Machines (GBM), AdaBoost, ExtraTrees, QLattice, and Bagging Regressor with Decision Trees. Shapley Additive Explanations (SHAP) were employed to interpret the black-box models. Model performance was evaluated using Mean Absolute Error (MAE), Root Mean Squared Error (RMSE), Coefficient of Determination (R^2), and computation time. The optimised GBM model, utilising six input parameters, demonstrated high accuracy, attaining a 0.55—30% higher MAE as compared to the PMV model. The findings underscore the potential of explainable, data-driven machine learning methods in capturing the influence of environmental, air quality, and personal factors on predictions of thermal comfort in air-conditioned buildings.

References:

Q. Chai, H. Wang, Y. Zhai, and L. Yang, “Using machine learning algorithms to predict occupants’ thermal comfort in naturally ventilated residential buildings,” Energy Build., vol. 217, p. 109937, 2020, doi: 10.1016/j.enbuild.2020.109937.

M. Luo et al., “Comparing machine learning algorithms in predicting thermal sensation using ASHRAE Comfort Database II,” Energy Build., vol. 210, p. 109776, 2020, doi: 10.1016/j.enbuild.2020.109776.

G. Zheng, Y. Zhang, X. Yue, and K. Li, “Interpretable prediction of thermal sensation for elderly people based on data sampling, machine learning and SHapley Additive exPlanations (SHAP),” Build. Environ., vol. 242, no. July, p. 110602, 2023, doi: 10.1016/j.buildenv.2023.110602.

Keyword: *GBM, Thermal Comfort, XAI*

Evaluation of Passenger s comfort Vote(PCV) and Discomfort Index(DI) for passengers travelling in non-air conditioned railway coach during summer

Shashank Shekhar Mishra[1], Vivek Kumar Gaba[2], Nisha Netam[3]
[1] Research Scholar, Department of Mechanical Engineering, National Institute of Technology Raipur, Chhattisgarh India
[2,3]Associate Professor, Department of Mechanical Engineering, National Institute of Technology Raipur, Chhattisgarh India
***Corresponding email:** nishane.me@nitrr.ac.in

ABSTRACT

Providing basic thermal comfort to public commuters is still considered a luxury in developing countries like India. Consequently, no significant design modification has been done in non-air conditioned railway coaches in past decades keeping thermal comfort in mind. In this context, Thermal comfort vote (TCV) was gathered from Indian railways' passengers from non-air-conditioned coach during the months of March and April 2022. A digital survey along with paper-based survey form was used to gather nearly 5000 datasets from the more than 300 participant passengers. Despite the fact that March and April are not the peak summer months in Chhattisgarh, the majority of the subjects reported feeling extremely uneasy. Globe and dry bulb temperature, relative humidity and air velocity were also measured inside the coach for evaluation of discomfort index(DCI) which is a function of psychological and environmental parameters. When compared to female passengers, the value of the discomfort index (DCI) for male passengers has been found to be 27% higher. On the other hand, the majority of passengers (63.4%) were satisfied with their humidity level. Most passengers also prefer their windows to be open (partially or completely) and their fans to be turned on. On the other hand, those who chose to turn off their fans in the afternoon were found to be happier with their thermal sensation. The findings indicate that thermal comfort while travelling is a highly subjective experience. Although the impact of the environment and climate on thermal comfort cannot be overstated, improving thermal comfort necessitates the evaluation of a variety of subjective and psychological parameters.

References:

1. Sansaniwal S. K. et. al. (2020), Review of studies on thermal comfort in Indian residential buildings, Science and Technology for the built environment,6-26727-748.
2. Zhang S., et. al.,(2020), Adaptive-rational thermal comfort model: Adaptive predicted mean vote with variable adaptive coefficient, Indoor Air,1–11.
1. Gao S, Ooka R, Oh W(2020) , Experimental investigation of the effect of clothing insulation on thermal comfort indices, Building and Environment

Keyword: *Thermal Comfort, Passenger's comfort vote (PCV), Discomfort index(DCI)*

INVESTIGATING THE EFFECT OF CANYON ASPECT RATIO ON THERMAL COMFORT THROUGH MUTUAL SHADING AND NATURAL VENTILATION POTENTIAL IN RESIDENTIAL URBAN BLOCKS OF WARM-HUMID CLIMATE

Milan Ben * , **Subbaiyan G**
Department of Architecture, National Institute of Technology, Tiruchirappalli

Thermal comfort studies are crucial for enhancing the quality of life, energy efficiency, and well-being in the built environment. The increasing reliance on mechanical systems for thermal comfort has led to higher energy demand, emphasizing the need for passive strategies at both building and urban scales. Urban morphology significantly influences thermal comfort, particularly in dense areas where factors like airflow and mutual shading are critical. Understanding the combined effects of mutual shading and natural ventilation is vital in warm and humid climates, where ventilation is essential due to excessive humidity.

This research investigates the impact of varying Canyon aspect ratios on thermal comfort in residential Urban Blocks in the climatic context of Chennai, Tamil Nadu, considering mutual shading and natural ventilation potential. A whole-building energy model was used to simulate indoor thermal environments in hypothetical urban blocks, based on Martin and March, with varying canyon aspect ratios. Additionally, Computational Fluid Dynamics simulations were conducted to analyse indoor airflow in the same units. The results from both models were integrated to evaluate indoor thermal comfort comprehensively.

Findings suggest that while indoor airspeed increases with a decrease in aspect ratio, the reduction in mutual shading leads to higher indoor air temperatures, which negatively impacts thermal comfort more significantly than the benefits of enhanced natural ventilation. Consequently, thermal comfort decreases as aspect ratios are reduced. Such a study is crucial in understanding the trade-offs among multiple influencing factors and how these collectively affect thermal comfort, helping prioritize the aspects that require more focus.

Keyword: *Indoor Thermal Comfort, CFD, Urban Morphology*

Investigating the effect of Canyon Aspect Ratio on Thermal Comfort through Mutual shading and Natural Ventilation Potential in Residential Urban Block of Warm-Humid Climate.

Milan Ben[1],*,G Subbaiyan[1]

[1] National Institute of Technology, Tiruchirappalli, India
* *401123002@nitt.edu*

SUMMARY
In developing countries like India, limited access to electricity and cooling makes indoor thermal comfort a challenge, especially amid rising global temperatures. Passive strategies like natural ventilation are crucial, particularly in humid cities like Chennai. However, urban-poor housing often adopts high-rise, high-density designs that affect ventilation and solar exposure. This study examines the impact of canyon aspect ratios on thermal comfort in low-income residential blocks in Chennai, considering mutual shading and natural ventilation. Hypothetical blocks with varying aspect ratios were analysed using Energy Plus for thermal performance and Ansys for airflow simulation. Results show that in summer, narrower canyons improve comfort in lower units due to shading, while upper units benefit from wider canyons that enhance air movement. These findings highlight the need to optimise block morphology to balance shading and ventilation for improved thermal comfort in dense urban housing.

KEYWORDS
Indoor Thermal Comfort, CFD, Operative temperature, Building Energy Model, Urban Canyon

1 INTRODUCTION

The average temperature around the world is increasing due to global warming, leading to a significantly higher frequency of periods of intense heat (Hansen et al., 2006). This makes it even more difficult to achieve thermally comfortable spaces passively, as well as increasing the energy demand required for cooling the space down (Santamouris, 2016). With the rising demand, another key aspect has to be dealt with to provide comfortable living spaces, that is Energy Poverty. Even though Space cooling is the most commonly used strategy for the mitigation of Heat, it is not economically feasible for many. In India, the rate of mortality compared between the heat wave period with non-heat wave period points towards an increase of 18.1 % (Nori-Sarma et al., 2019), and this could be potentially linked to their thermally uncomfortable living environments. Hence, this stresses the need for passive approaches, alternative to Air conditioning, to improve indoor comfort in living spaces and reduce the health risks associated with rising global temperatures (Jiang et al., 2023).

1.1 Natural Ventilation As A Passive Strategy

The Hot and Humid tropical belt contains many developing nations that are highly populated, covering up to 70 per cent of the World Population (Kaushik, 1988). High humidity poses a challenge to thermal comfort, and with dehumidification and cooling often unaffordable in developing countries like India, enhancing natural ventilation emerges as the most effective passive strategy. It also helps lower cooling energy demand and related greenhouse gas emissions. The study done by (Stasi et al., 2024) found that natural ventilation significantly enhanced indoor thermal conditions in low-income housing in hot, arid climates. Cross-ventilation reduced indoor temperatures by up to 3.1°C on the hottest day and 3.9°C during typical cooling days, while discomfort hours dropped by 31.3% and 77.8%, respectively.

1.2 Natural Ventilation And Urban Block Characteristics

Natural ventilation potential is strongly influenced by the surrounding context, as urban canyon geometry can alter wind flow and thermal buoyancy-driven ventilation (Syrios & Hunt, 2007). The study indicates that airflow patterns and Air Changes per Hour (ACH) vary significantly between narrow and wide canyons (Syrios & Hunt, 2007). Hence, it is important to consider the effect of the Neighbouring geometry when doing a Natural Ventilation study.

1.3 Impact Of Canyon Aspect Ratio On Heat Gain And Cooling Load

Urban canyon geometry, especially its aspect ratio (building height to street width), strongly affects solar access to façades and impacts both outdoor and indoor thermal conditions (Ali-Toudert & Mayer, 2006). In narrow canyons with high aspect ratios, increased mutual shading reduces solar exposure on façades, lowering heat gain, indoor temperatures, and cooling energy demand (Krüger et al., 2010).

1.4 Aim Of The Study

This research investigates the impact of Canyon aspect ratios on thermal comfort in low-income residential blocks of Chennai, Tamil Nadu, considering mutual shading and natural ventilation potential.

- To perform radiation analysis to assess the effect of mutual shading with variations in canyon aspect ratio.
- To evaluate the effect of changes in canyon aspect ratio on Indoor air speed.
- To compute the comfort hours based on indoor air temperature and airspeed for each variation in canyon aspect ratio.

2 METHODOLOGY

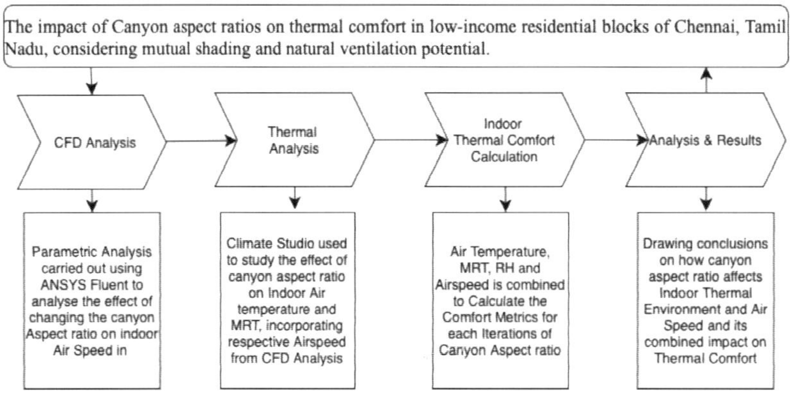

Figure 1. Methodology Chart

2.1 Study Area Description

The study area is located in Perumbakkam, Chennai, at 12.92° N, 80.24° E. Chennai, on the south-eastern coast of India, has a tropical wet and dry climate ("Aw" under Köppen classification). It experiences high temperatures year-round, ranging from 25°C to 35°C, with the hottest months (April to June) exceeding 40°C. The Predominant Wind Direction at 1000 hPa is from South to North, followed by South-West to North-East (IMD, 2020) as shown in Figure 2.

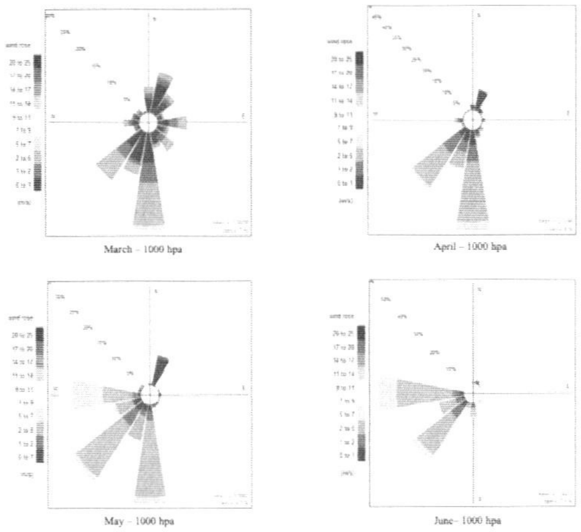

Figure 2. Wind Rose Diagram of Chennai, Source - (IMD, 2020)

The study area is a settlement developed by the Tamil Nadu Slum Clearance Board (TNSCB), housing over 30,000 households with various residential typologies. The predominantly low-income residents lack access to space cooling or forced ventilation systems, relying entirely on passive strategies, which makes it ideal for studying the effectiveness of natural ventilation.

2.2 Shoebox Model And Iterations

This block type, adapted to the Perumbakkam context from Martin and March's Urban Building Typologies, is referred to as Street/Slab. The urban block consists of three residential buildings, each with a ground floor plus seven additional floors (G+7), as shown in Figure 3. Each building measures 62m by 12.2m at the base.

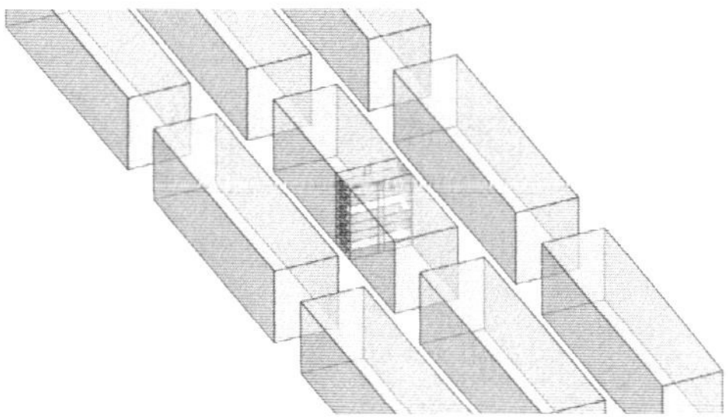

Figure 3. Shoebox model for CFD and Thermal Analysis

The blocks are oriented east-west, perpendicular to Chennai's prevailing summer winds. The central building is analysed with two target apartments (G+2 and G+6) modelled with open windows for CFD and thermal analysis. 4 canyon widths 12m, 18m, 24m, and 30m are tested.

Table 1. Urban Block Characteristics

No of Buildings	3
Street widths (m) Iterations	12, 18, 24, 30
No of Floors	G+7
Building Height (m)	24 m
Building Type	Residential

2.3 CFD Simulation

The behaviour of wind-driven natural indoor ventilation was simulated to assess the average air velocity within the building (centre, at 1.1m height). The Three-Dimensional Geometry of the Urban Block was modelled using Ansys Discovery SpaceClaim. The computational fluid dynamics (CFD) Outer domain used for this analysis is as suggested by (Mei et al., 2017). The height of the building (h) serves as the base for the dimensions of the larger domain. The upstream, downstream, lateral, and height of the larger domain (H) are 13.3h, 28h, 10.8h and 7h, respectively as shown in Figure 4.

For the numerical study, a tetrahedral mesh was generated with a grid size of 5m in the domain and a finer and refined 1m around the openings and the interior of the buildings for accurate results. A wind velocity of 2.3 m/s obtained from the weather data of Chennai, measured at a height of 12 m above the ground, was used as the inlet boundary condition for the outer domain. The inlet velocity was applied using the Weibull function :

$Vh=Vr\ (h/hr)\alpha$, α=0.27 *is the terrain* roughness for suburban area (Izadyar et al., 2023)

where *Vh* is the velocity inlet profile related to a specific height, and *Vr* & *hr* are the velocity and height of the MET Station, as the simulation reference.

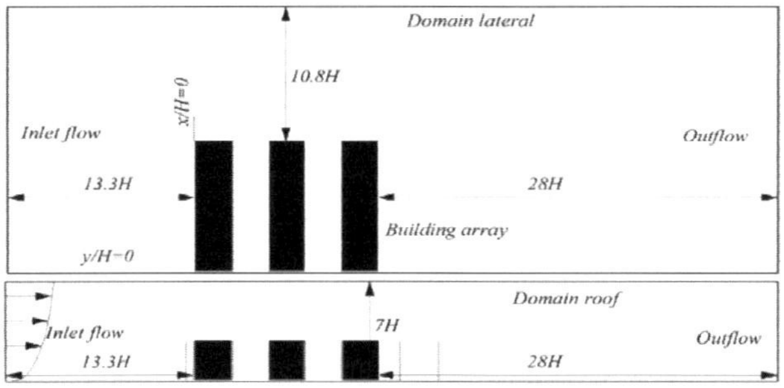

Figure 4. Schematic Diagram of Computational Domain. The top Figure is the Top View and the bottom Figure is the Side View, where H is the height of the buildings.

2.4 Thermal Model

To assess the impact of canyon aspect ratio on thermal performance, an hourly dynamic thermal simulation was carried out using Climate Studio. Indoor ventilation data from CFD analysis was integrated into the building's thermal model, following the approach used by (Stasi et al., 2024). The EnergyPlus Weather file for Chennai was used as the Input Weather data.

Table 2. Input parameters– Climate Studio (Energy Plus)

Wall Assembly U-Value	2.28 W/(m²K)
Roof U-Value	3.59 W/(m²K)
WWR (Window-to-Wall Ratio)	0.25
Fenestration U-Value	3.5 W/(m²K)
Fenestration SHGC	0.68
Fenestration T.vis	0.81
Run Period	1 March – 30 June

Air Temperature and Mean Radiant Temperature data were extracted for 2 target apartments (G+2 & G+6) from the Simulation results of each iteration of the Canyon Aspect Ratio

2.5 Thermal Comfort Analysis

The air temperature (Ta) and mean radiant temperature (Tmrt) obtained from the thermal model, along with the airspeed inside the apartment, are used to calculate the Operative Temperature (Stasi et al., 2024) for all 8760 hours. This Operative Temperature is then compared with the IMAC Comfort Band for naturally ventilated buildings to determine the number of hours that fall within the comfortable range for each iteration of the canyon aspect ratio. The equation to calculate operative temperature:

$$Top = (Tmrt + (Ta \times \sqrt{10v})) / (1 + \sqrt{10v}), \qquad when\ v > 0.20\ m/s$$

Where,

Top is the Operative Temperature, *Tmrt* is the Mean Radiant temperature, *Ta* is the Air Temperature, and *V* is the Air Velocity in m/s (Sudarsanam & Kannamma, 2023).

3 RESULTS AND DISCUSSION

3.1 CFD Analysis

The CFD analysis shows very distinct results for upper (G+6) and lower (G+2) units, pointing towards the significant impact of the canyon width on Indoor Natural ventilation potential as shown in Figure 5.

The Lower Apartment unit showed increased indoor air velocity as canyon width increased from 12 m to 18 m, then dropped significantly at 24 m. In contrast, the Upper unit exhibited lower air speeds and an opposite trend. This difference is likely due to pressure variations caused by "Skimming Flow," where minimal airflow penetrates the canyon canopy (Makedonas et al., 2021). The airflow direction in the apartments was found to be opposite to the prevailing wind. This wind reversal in urban canyons can be attributed to higher surface pressure on the leeward façade compared to the windward façade of the ventilated building (Syrios & Hunt, 2007)

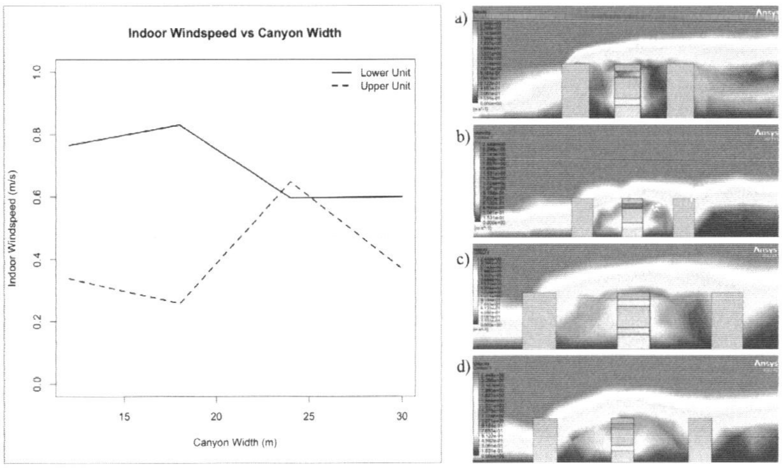

Figure 5. Indoor Wind Speed plotted against canyon width for 2 Apartment Units alongside Wind Velocity Contours of different canyon widths- a) 12m, b)18m, c) 24m, d) 30m.

3.2 Thermal Comfort Analysis

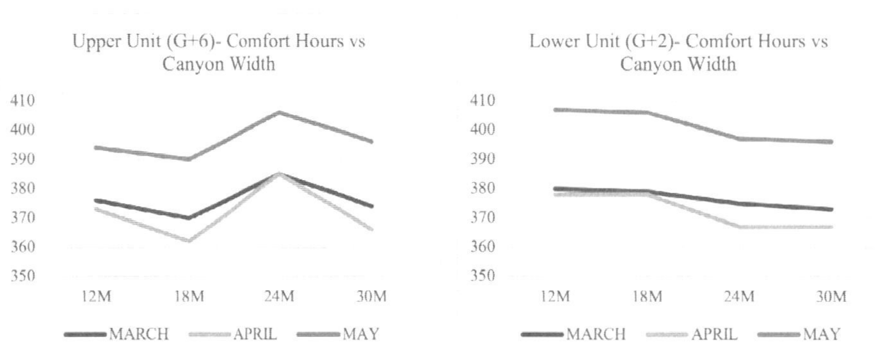

Figure 6. Variation of Comfortable Hours in the Lower Unit (Right) and Upper unit (Left) with Increasing Canyon Width Across Months

Operative temperature for the two target apartments (G+2 & G+6) was calculated hourly for the summer months using thermal model data (indoor air and mean radiant temperature) and CFD-derived air velocity. Comfort hours were then assessed using IMAC data for March to June. The lower unit showed a steady drop in comfort hours with increasing canyon width (Figure 6), due to higher direct radiation exposure from larger canyon aspect ratios (Figure 7).

In the case of the Upper Unit (G+6), the number of comfortable hours exhibits a non-linear trend with increasing canyon width. As shown in Figure 6, comfort hours initially decrease when the canyon width increases to 18 meters but rise again as the width increases to 24 meters. This behaviour can be attributed to the fact that the Upper Unit is already significantly exposed to direct radiation in the base case. Consequently, further increases in canyon width do not result in a proportionally large rise in radiation exposure (Figure 7). This explains the relatively increasing trend in comfort hours for the Upper Unit, in contrast to the consistent decline observed in the Lower Unit.

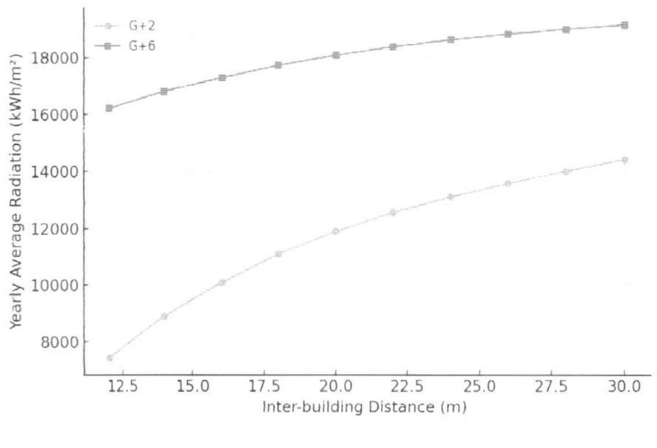

Figure 7: Variation in Direct Radiation Exposure on Units with Increasing Canyon Width

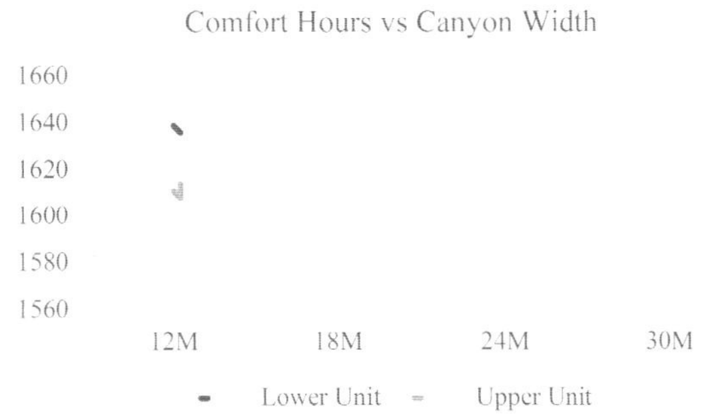

Figure 8: Comparison of Total Summer Comfort Hours for Lower and Upper Units Across Varying Canyon Widths

When considering the total number of comfort hours during the summer months, the lower unit performs better in narrow canyon configurations, while the upper unit records higher comfort hours in wider canyons, as illustrated in Figure 8.

This contrast can be attributed to the increased exposure of the lower unit to direct radiation and reduced indoor air speed in wider canyons, which negatively impacts thermal comfort. Conversely, in the upper unit, wider canyons facilitate greater indoor air movement, which helps lower the operative temperature, thereby enhancing thermal comfort levels.

4 CONCLUSION

In conclusion, the analysis of thermal comfort across varying canyon widths reveals distinct trends for both the lower and upper units. The lower unit experiences a decline in comfort hours with increased canyon width, primarily due to greater exposure to direct solar radiation and reduced shading. In contrast, the upper unit shows a more complex response, with comfort hours initially decreasing at 18 meters of canyon width but increasing at wider configurations (24 meters), owing to improved indoor air movement that helps regulate operative temperatures. Overall, narrower canyons offer better comfort for the lower unit, while wider canyons enhance comfort for the upper unit, highlighting the importance of canyon configuration in shaping thermal comfort in urban environments. The Study also reveals that the canyon aspect ratio of 1.1 provides optimal thermal conditions for both the lower and upper units. Therefore, a canyon aspect ratio of 1.1 emerges as the most effective configuration for maintaining comfortable thermal conditions in both units across the studied scenarios.

Further studies on different block configurations, building heights, and shapes are needed to better understand their impact on thermal comfort. Exploring a broader range of urban forms will provide deeper insights into how these factors interact with canyon width to influence both solar radiation exposure and indoor air movement, ultimately informing more effective design strategies for optimising comfort in diverse urban contexts.

5 REFERENCES

Ali-Toudert, F., & Mayer, H. (2006). Numerical study on the effects of aspect ratio and orientation of an urban street canyon on outdoor thermal comfort in hot and dry climate. *Building and Environment*, *41*(2), 94–108. https://doi.org/10.1016/j.buildenv.2005.01.013

Hansen, J., Sato, M., Ruedy, R., Lo, K., Lea, D. W., & Medina-Elizade, M. (2006). Global temperature change. *Proceedings of the National Academy of Sciences*, *103*(39), 14288–14293. https://doi.org/10.1073/pnas.0606291103

IMD. (2020). *Monthly Upper Air Windrose Diagrams (2000-2015)- Chennai*. IMD. https://imdpune.gov.in/library/public/CHE_43279_L.pdf

Izadyar, N., Miller, W., Rismanchi, B., Garcia-Hansen, V., & Matour, S. (2023). Balcony design and surrounding constructions effects on natural ventilation performance and thermal comfort using CFD simulation: A case study. *Journal of Building Performance Simulation*, *16*(5), 537–556. https://doi.org/10.1080/19401493.2023.2185682

Jiang, Z., Kobayashi, T., Yamanaka, T., & Sandberg, M. (2023). A literature review of cross ventilation in buildings. *Energy and Buildings*, *291*, 113143. https://doi.org/10.1016/j.enbuild.2023.113143

Krüger, E., Pearlmutter, D., & Rasia, F. (2010). Evaluating the impact of canyon geometry and orientation on cooling loads in a high-mass building in a hot dry environment. *Applied Energy*, *87*(6), 2068–2078. https://doi.org/10.1016/j.apenergy.2009.11.034

Makedonas, A., Carpentieri, M., & Placidi, M. (2021). Urban Boundary Layers Over Dense and Tall Canopies. *Boundary-Layer Meteorology*, *181*(1), 73–93. https://doi.org/10.1007/s10546-021-00635-z

Mei, S.-J., Hu, J.-T., Liu, D., Zhao, F.-Y., Li, Y., Wang, Y., & Wang, H.-Q. (2017). Wind driven natural ventilation in the idealized building block arrays with multiple urban morphologies and unique package building density. *Energy and Buildings*, *155*, 324–338. https://doi.org/10.1016/j.enbuild.2017.09.019

Nori-Sarma, A., Anderson, G. B., Rajiva, A., ShahAzhar, G., Gupta, P., Pednekar, M. S., Son, J.-Y., Peng, R. D., & Bell, M. L. (2019). The impact of heat waves on mortality in Northwest India. *Environmental Research*, *176*, 108546. https://doi.org/10.1016/j.envres.2019.108546

Santamouris, M. (2016). Cooling the buildings – past, present and future. *Energy and Buildings*, *128*, 617–638. https://doi.org/10.1016/j.enbuild.2016.07.034

Stasi, R., Ruggiero, F., & Berardi, U. (2024). Natural ventilation effectiveness in low-income housing to challenge energy poverty. *Energy and Buildings*, *304*, 113836. https://doi.org/10.1016/j.enbuild.2023.113836

Sudarsanam, N., & Kannamma, D. (2023). Investigation of summertime thermal comfort at the residences of elderly people in the warm and humid climate of India. *Energy and Buildings*, *291*, 113151. https://doi.org/10.1016/j.enbuild.2023.113151

Syrios, K., & Hunt, G. R. (2007). Urban Canyon Influence on Building Natural Ventilation. *International Journal of Ventilation*, *6*(1), 43–49. https://doi.org/10.1080/14733315.2007.11683763

Assessment of occupant's thermal comfort and performance in naturally ventilated buildings

Virendra Sharma * [1], **Faizan Ahmad Wani[2], Jyotirmay Mathur[1], Sanjay Mathur[2]**

[1] *School of Energy and Clean Technology, NICMAR University of Construction Studies, Hyderabad, India. 500101*

[2] *Department of Civil Engineering, Malaviya National Institute of Technology Jaipur, India. 302017*

In naturally ventilated buildings, the thermal environment and personal factors change with the seasons, directly impacting occupant thermal comfort and performance. Investigating the combined effect of these factors on thermal comfort and performance and the interrelationship between thermal comfort and performance is vital for deeper insights. This research evaluates thermal comfort, performance, mental demand and frustration based on the Standard Effective Temperature (SET), a common index that integrates thermal environmental and personal factors. Forty subjects with equal male and female ratio participated in this experiment, which was conducted once a month from January to April in naturally ventilated buildings. Thermal comfort, mental demand, and frustration were assessed through subjective questionnaires, with the ASHRAE 55 scale used for thermal comfort and NASA TLX for mental demand and frustration. Seven neurobehavioral tests were used to quantify overall performance, categorized into four groups: attention (Visual Choice Reaction Time), perception (Stroop Test, Picture Recognition, Graph Overlapping), memory (Digit Span, Digit Symbol Substitution), and higher-order cognitive skills (Serial Addition and Subtraction). The results reveal that "Slightly cool to Neutral" thermal sensation in naturally ventilated buildings provides better thermal comfort and performance and minimizes mental demand and frustration. Furthermore, the findings indicate that an occupant-perceived temperature, i.e. SET in the range of 23°C to 25°C, yields better thermal comfort and performance and minimizes mental demand and frustration. Occupants in naturally ventilated buildings adapt to maintain their SET within this range by altering clothing in winter and increasing air velocity during summer. These adaptive behaviours effectively support the body's thermoregulation, ensuring thermal comfort and improving performance. The outcome of this research emphasizes the need for climate-responsive design strategies to improve the indoor thermal environment and promote the adaptive behaviours strategy to bring their SET within the range for better thermal comfort and performance in naturally ventilated buildings.

Keyword: *Thermal comfort, Performance, Standard effective temperature*

Impact of Building Orientation and Height on Thermal Conditions in Malaysian School Classrooms: A Field Study in Kuala Lumpur

Nor Sahidah Firman[1], Sheikh Ahmad Zaki * [1], Ng Wai Tuck[2], Manoj Kumar Singh[3], Hom Bahadur Rijal[4], Samsol Faizal Anis[1]

[1] *Malaysia-Japan International Institute of Technology, Universiti Teknologi Malaysia, Jalan Sultan Yahya Petra, 54100 Kuala Lumpur, Malaysia*

[2] *Faculty of Built Environment, Tunku Abdul Rahman University of Management and Technology, Jalan Genting Kelang, 53300 Kuala Lumpur, Malaysia*

[3] *Energy and Sustainable Built-Environment Design (ESBD) Laboratory, Department of Civil Engineering, Shiv Nadar Institution of Eminence, Greater Noida, Uttar Pradesh, 201314, India*

[4] *Faculty of Environmental Studies, Tokyo City University, Yokohama 224-8551, Japan*

Building-related factors such as orientation and height are among several passive elements influencing indoor air temperature. To investigate these factors, thirty-two classrooms from eight secondary schools in Kuala Lumpur were selected for field study. The measurements were conducted throughout one school day simultaneously in four classrooms at each school, to record the thermal environment. The analysis of variance found that classroom orientation has no significant overall impact on indoor air temperature. The high indoor air temperatures were observed mostly in classrooms near or below the roof. In one school, the indoor temperature difference between classrooms with a similar orientation, located in different building blocks and floor levels, one on the second floor (below the roof) and the other on the third floor, was around 1.4 °C. Exploring passive design strategies to enhance comfort in school buildings is necessary in tropical climates, focusing on minimizing the impact of external heat.

Keyword: *Indoor temperature, Passive elements, Thermal comfort*

Occupational Impact on Thermal Comfort Experiences in Rural Environments

Chaman Lal Pal[1], Nisha Netam[2*]
[1] Research Scholar, Department of Mechanical Engineering, National Institute of Technology Raipur, Chhattisgarh India
[2]Associate Professor, Department of Mechanical Engineering, National Institute of Technology Raipur, Chhattisgarh India
*Corresponding email: nishane.me@nitrr.ac.in

This study investigates occupation-based thermal comfort predictions in rural areas. While previous research has explored the influence of age, gender, and body mass index (BMI) on thermal comfort, the impact of occupational nature remains underexplored. To address this gap, data were collected from 511 and 497 respondents during the summer and winter seasons, respectively, using questionnaire surveys, observations, and environmental measurements. We applied a linear regression model to determine thermal comfort temperatures for different occupational groups.

The winter season coincides with peak agricultural activities, allowing for a comparison of thermal adaptation across three occupational categories: educational, agricultural, and homemaker. Results indicate that agricultural workers experience higher thermal comfort temperatures than other groups during the summer. Conversely, during winter, individuals in the educational group report feeling cooler than those in other occupations.

These results suggest that metabolic rate might not fully explain changes in thermal comfort, since long-term exposure at work causes different thermal adaptations. This study highlights that the nature of work and environmental exposure significantly influences variations in thermal comfort temperatures. Incorporating these factors into thermal comfort standards can improve the accuracy and inclusivity of design guidelines for built environments.

References:

M. Indraganti, R. Ooka, H.B. Rijal, Thermal comfort in offices in India: Behavioral adaptation and the effect of age and gender, Energy Build. 103 (2015) 284–295.

Pal, C. L., Netam, N., Sanyal, S., & Manish, M. (2024). Exploring thermal comfort in rural houses of Chhattisgarh state, India: A comprehensive survey and adaptive model analysis. Science and Technology for the Built Environment, 30(5), 523-545.

Aqilah, N., Rijal, H. B., & Yoshida, K. (2023). Gender difference regarding thermal comfort in Japanese residential building during free running mode. Building and Environment, 245, 110891.

Keyword: *thermal comfort, rural areas, occupational*

Thermal Performance of Different Types of Green Roofs in Nepalese Residential Buildings

Alfa Shrestha Pradhan[1], Prativa Lamsal[1], Hom Bahadur Rijal[2], Samjhana Bhattarai[1], Nira Shrestha[1], [1]Tribhuvan University, [2]Department of Restoration Ecology & Built Environment

Urbanization has globally led to the urban heat island (UHI) effect, raising temperatures, energy use, and reducing thermal comfort. Kathmandu Valley, a rapidly urbanizing region in South Asia, faces rising temperatures and growing energy demands due to unplanned development. Green roofs have emerged as a solution to reduce temperatures through the insulating properties of vegetation and soil, enhancing urban livability. Although a new concept in Nepal, they have primarily been explored for aesthetics, with thermal benefits largely unexplored. This study aims to assess the thermal performance of three green roof types—intensive (deep substrata), extensive (shallow substrata), and rooftop farming (cultivation)—in comparison to traditional RCC roofs on residential buildings in Kathmandu Valley. The research focuses on three randomly selected residential buildings, each featuring one type of green roof on the lower terrace and an RCC roof on the upper terrace. Thermal performance was analyzed by continuously measuring indoor and outdoor air temperatures, with sensors placed 1.5 meters above the floor level and 0.3 meters above the green roof level, over a ten-day period during the peak summer months. The results show that indoor air temperatures were reduced by 2.4°C, 1.6°C, and 0.8°C for intensive, extensive, and rooftop farming roofs, respectively, compared to RCC roofs. Additionally, outdoor air temperature around the roof surface were lower by 0.6°C for intensive and extensive roofs, and 0.4°C for rooftop farming roofs. This study shows that intensive green roofs outperform RCC roofs in thermal mitigation, reducing temperatures and offering a viable solution to the UHI effect.

Keywords: Intensive green roof, Extensive green roof,Rooftop farming, Air temperatureNepal, Kathmandu valley, Residential building

Ambient Air Quality and Air Cleaning Strategies in India

Christopher Muller *

AAF International, Louisville, Kentucky, USA

India continues to grapple with critical air pollution challenges, particularly in urban areas where rapid industrialization, vehicular emissions, and agricultural practices have severely impacted air quality. India is one of the five most polluted countries globally and has six of the ten most polluted cities; Byrnihat, Delhi, Mullanpur, Faridabad, Loni, and New Delhi (IQAir, 2025). The major pollutants include particulate matter ($PM_{2.5}$ and PM_{10}), nitrogen oxides (NOx), sulfur dioxide (SO_2), carbon monoxide (CO), and volatile organic compounds (VOCs), with particulate pollution being a leading concern. This presentation examines the sources of air pollution in India, its implications on public health and the environment, and explores potential air cleaning strategies, including policy interventions, technological innovations, and community-driven initiatives.

References:

Central Pollution Control Board (CPCB). (2020). National Ambient Air Quality Status & Trends 2019-2020. Retrieved from http://cpcb.nic.in.

IQAir. (2025). 2024 World Air Quality Report.

Keyword: *Air pollution, air quality, air cleaning strategies*

Ambient Air Quality and Air Cleaning Strategies in India

Christopher Muller

AAF International, Louisville, Kentucky, USA
Corresponding email: christopher.muller@aafintl.com

Abstract
India continues to grapple with critical air pollution challenges, particularly in urban areas where rapid industrialization, vehicular emissions, and agricultural practices have severely impacted air quality. India is one of the five most polluted countries globally and has six of the ten most polluted cities; Byrnihat, Delhi, Mullanpur, Faridabad, Loni, and New Delhi [1]. The major pollutants include particulate matter ($PM_{2.5}$ and PM_{10}), nitrogen oxides (NOx), ozone (O_3), sulfur dioxide (SO_2), carbon monoxide (CO), and volatile organic compounds (VOCs), with particulate pollution being a leading concern. Enhanced air cleaning systems that integrate advanced filtration technologies are essential to maintaining good indoor air quality (IAQ), especially in highly polluted urban environments. This paper examines the sources of air pollution in India, its implications on public health and the environment, and explores potential air cleaning strategies, including policy interventions, technological innovations, and community-driven initiatives [2,3].

Keywords: Air pollution, India, air quality, public health, air cleaning strategies, NCAP [4], *vehicular emissions*

1. Introduction

India is currently one of the countries most affected by poor air quality, both outdoors and indoors. The country relies heavily on coal for electricity generation, biomass for cooking in rural areas, and is experiencing rapid growth in vehicular traffic and industrial activity. These contribute to the high emissions of $PM_{2.5}$ and PM_{10}, NO_2, SO_2, O_3, and volatile organic compounds (VOCs). Motor vehicle emissions are the fastest growing new source of pollution in cities, and industry contributes many VOCs.

A 2023 report by the Centre for Science and Environment [5] notes that air pollution is responsible for more than 1.6 million premature deaths annually in India. According to the Central Pollution Control Board [6], 132 Indian cities are currently non-compliant with National Ambient Air Quality Standards [7], with $PM_{2.5}$ levels frequently exceeding safe limits.

India's annual average $PM_{2.5}$ pollution increased steadily between 1998 and 2020, with Delhi consistently among the most polluted regions. Over that period, the share of population meeting the national standard of 40 µg/m³ dropped from about 60% to ~28%. The historically most polluted Indian cities (annual AQI >300 in multiple months) often include Delhi, Ghaziabad, Noida, Kanpur, Lucknow, Patna, Varanasi, Agra, Muzaffarpur, and Gurugram [8,9]. Fig. 1 shows India's AQI category distribution along with the 10 most polluted Indian cities.

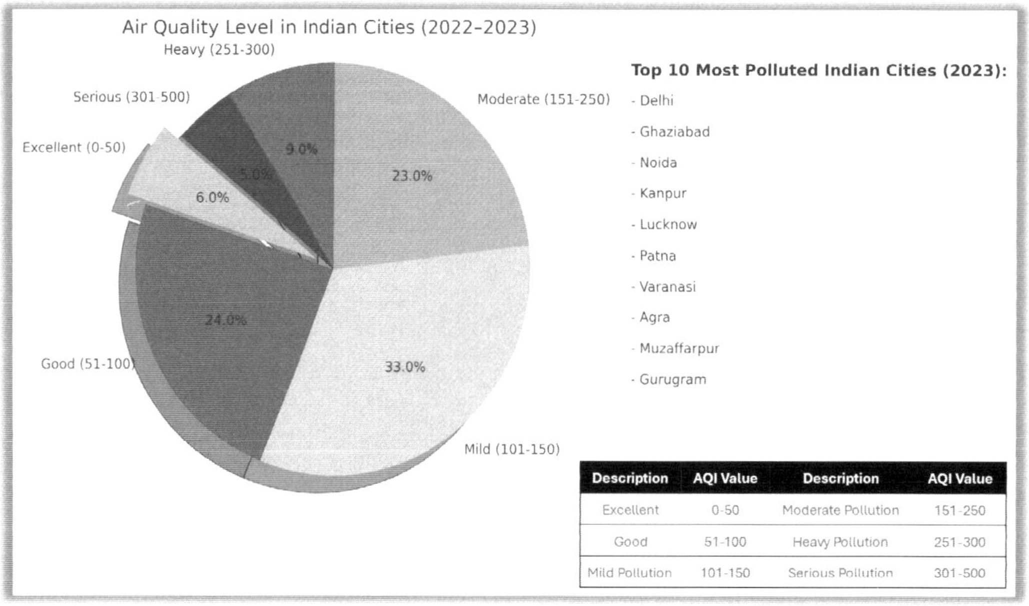

Figure 1. Air quality level in Indian cities (2022–2023) showing distribution by AQI category, top 10 most polluted cities, and AQI category ranges.

India operates an extensive network of air quality monitoring stations, covering over 550 cities and towns through the National Air Quality Monitoring Program (NAMP) [10], and the expanding network of Continuous Ambient Air Quality Monitoring Stations (CAAQMS) [11], managed by the Central Pollution Control Board (CPCB) [12]. A summary of the NAMP is provided in Table 1.

Table 1. India Air Quality Monitoring and Indoor Exposure

Metric	India (Current Status)
Cities with permanent monitors	~550 cities with stations; 1,524 total; most cities have only 1 station; ~15 cities have ≥5 stations
Monitoring coverage types	Mix of capitals and Tier-2/3 cities; wide geographic spread, lower station density
Pollution trends	In Delhi and other major cities, all monitored pollutants have upward long-term trends
Indoor formaldehyde exposures	Local evidence shows frequent exceedance; no national survey, but WHO guidelines exceeded in many urban households [13]

This monitoring system includes stations in all state and union territory capitals, as well as in major industrial, commercial, and residential areas. As of 2023, India has more than 1,500 monitoring stations, including both manual and continuous stations, with many located in critically polluted or highly urbanized regions such as the Indo-Gangetic Plain, Delhi-NCR, and other major metropolitan clusters. Although India's CAAQMS monitor pollutants in major cities, there continue to be data gaps and inconsistent monitoring coverage in smaller towns.

Analysis of long-term trends indicates that a significant share of these monitoring locations show persistent or increasing levels of air pollution, particularly for particulate matter ($PM_{2.5}$ and PM_{10}) and nitrogen dioxide (NO_2), reflecting growing urbanization, traffic density, and regional emission sources. Figure 2 shows satellite-based and reanalysis maps that show $PM_{2.5}$ levels across India, highlighting severe air pollution in the Indo-Gangetic Plain from 1998 to 2020 [14].

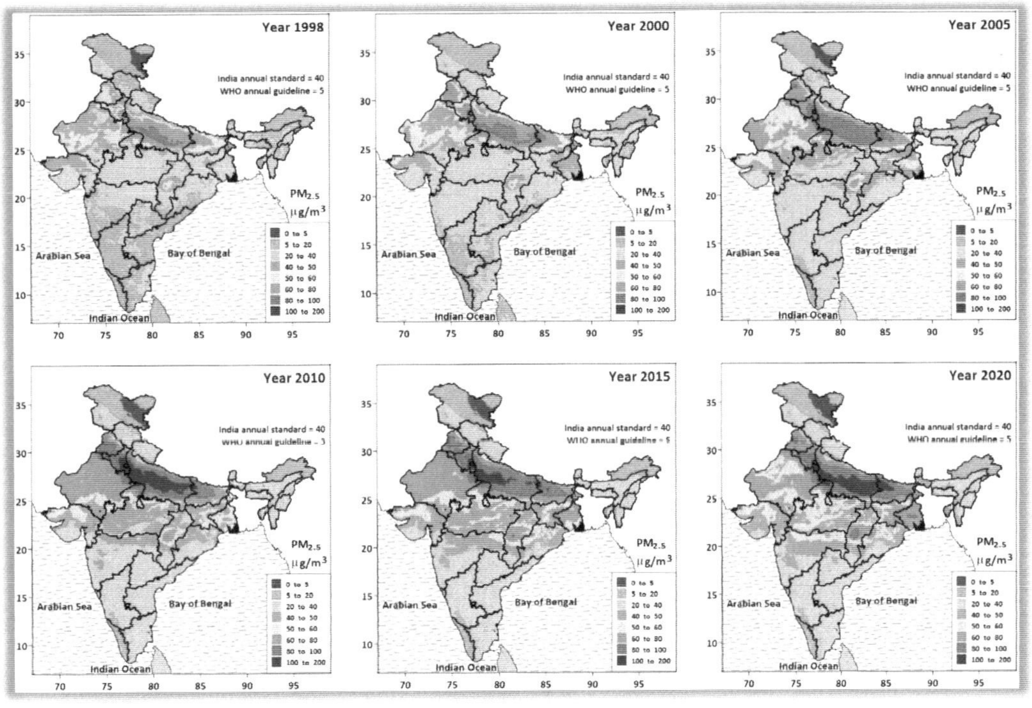

Figure 2. Evolution of $PM_{2.5}$ pollution across India (1998–2020).

Figure 3 shows the distribution of 2010 surface annual average $PM_{2.5}$ concentrations across India, by district quintiles, depicting higher pollution in the Indo-Gangetic Plain and north-central India [15].

The ambient air quality standards established by the Central Pollution Control Board (CPCB) under the National Ambient Air Quality Standards [6] differ significantly from the World Health Organization's Air Quality Guidelines [13]. Table 2 presents a comparison of key pollutant limits. The WHO guidelines are considerably more stringent for particulate matter, with recommended annual average concentrations of 15 µg/m³ for PM_{10} and 5 µg/m³ for $PM_{2.5}$, compared to India's limits of 60 µg/m³ and 40 µg/m³, respectively. Similarly, WHO's short-term (24-hour) values for $PM_{2.5}$ (15 µg/m³) and PM_{10} (45 µg/m³) are less than one-third of India's permissible levels.

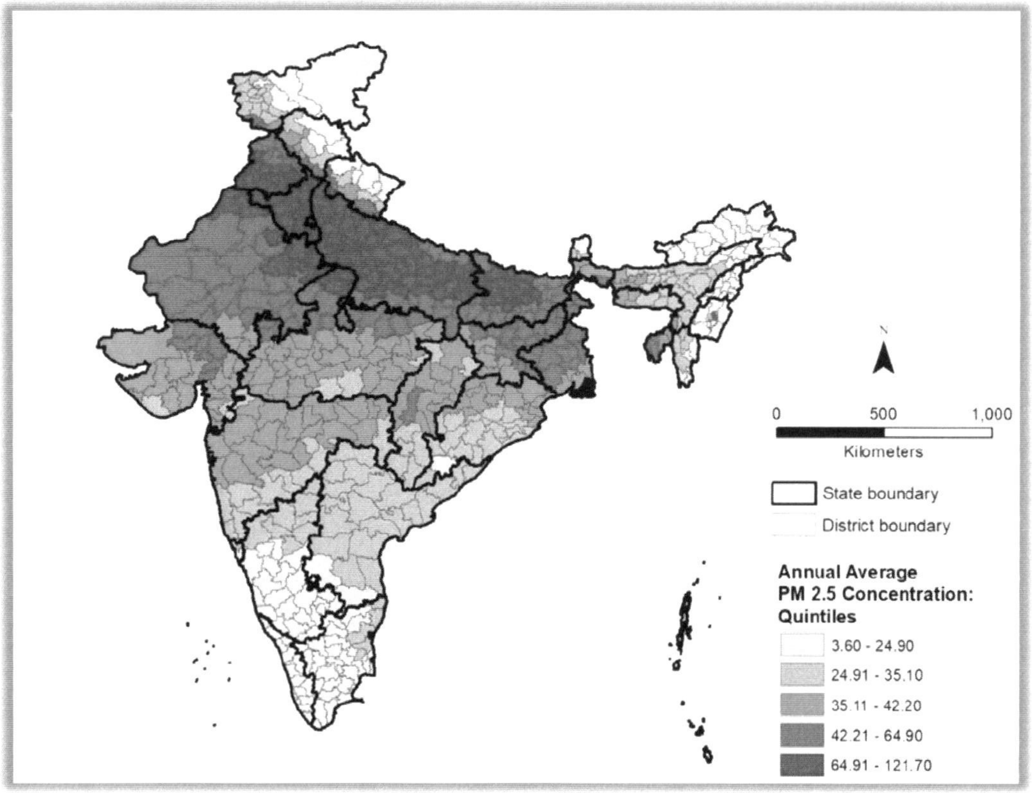

Fig. 3. Annual average $PM_{2.5}$ distribution across India in 2010.

Table 2. Comparison of ambient air quality standards for India and WHO guidelines

Pollutant	Averaging Time	Unit	India NAAQS (2009)	WHO AQG (2021)
PM_{10}	Annual	µg/m³	60	15
	24-hour	µg/m³	100	45
$PM_{2.5}$	Annual	µg/m³	40	5
	24-hour	µg/m³	60	15
NO_2	Annual	µg/m³	40	10
	24-hour	µg/m³	80	25
SO_2	Annual	µg/m³	50	40
	24-hour	µg/m³	80	40
O_3	8-hour	µg/m³	100	100
CO	8-hour	mg/m³	2	4
	1-hour	mg/m³	4	30

For nitrogen dioxide (NO_2), WHO's annual guideline of 10 µg/m^3 is one-quarter of India's 40 µg/m^3 standard, while the 24-hour NO_2 limit (25 µg/m^3) is substantially lower than India's 80 µg/m^3. Sulfur dioxide (SO_2) limits are also stricter in WHO guidelines (40 µg/m^3 annual; 40 µg/m^3 24-hour) compared to India's 50 µg/m^3 and 80 µg/m^3. For ozone (O_3), WHO recommends an 8-hour mean of 100 µg/m^3, which matches India's standard, indicating some alignment for this pollutant. In the case of carbon monoxide (CO), WHO prescribes an 8-hour value of 4 mg/m^3 and a 1-hour limit of 30 mg/m^3, while India's limits are set lower at 2 mg/m^3 (8-hour) and 4 mg/m^3 (1-hour), reflecting different risk management strategies.

Overall, the WHO guidelines represent health-based targets aimed at minimizing risks, while India's standards, though improved compared to earlier iterations, remain higher for most pollutants, particularly for particulate matter, indicating the potential for further tightening to align with global best practices.

2. India's Air Quality Index

India's Central Pollution Control Board (CPCB), under the Ministry of Environment, Forest and Climate Change (MoEFCC) [16], maintains comprehensive databases for ambient air pollutants across major cities and urban areas in India. The CPCB operates the National Air Quality Monitoring Program (NAMP) and an expanding network of Continuous Ambient Air Quality Monitoring Stations (CAAQMS), which provide near real-time measurements of key pollutants.

Daily reports of ambient air quality are published for hundreds of Indian cities, including the Air Quality Index (AQI), identification of primary pollutants, and the corresponding air quality category. Public access to current AQI data is available through the CPCB online portal [17].

The Indian AQI is a composite indicator derived from measurements of multiple contaminants, including $PM_{2.5}$, PM_{10}, SO_2, NO_2, CO, and O_3. From the AQI values, it is possible to interpret the relative contribution of individual pollutants and determine average pollutant concentration levels for different regions and time periods (Table 3).

Table 3. India AQI values and corresponding pollutant concentrations [18]

AQI	$PM_{2.5}$ ($\mu g/m^3$)	PM_{10} ($\mu g/m^3$)	NO_2 ($\mu g/m^3$)	SO_2 ($\mu g/m^3$)	CO ($\mu g/m^3$)	O_3 ($\mu g/m^3$)
0–50	0–30	0–50	0–40	0–40	0–1.0	0–50
51–100	31–60	51–100	41–80	41–80	1.1–2.0	51–100
101–200	61–90	101–250	81–180	81–380	2.1–10	101–168
201–300	91–120	251–350	181–280	381–800	10–17	169–208
301–400	121–250	351–430	281–400	801–1600	17–34	209–748
401–500	>250	>430	>400	>1600	>34	>748

3. Air Quality Data: Delhi, Mumbai, Kolkata

As expected, higher AQI levels are most often reported in large metropolitan areas of India, and the values correlate well with diminished air quality measured at Central Pollution Control Board (CPCB) monitoring stations [19,20]. For this paper, the investigation focused on air quality in Delhi, Mumbai, and Kolkata, with a primary interest in SO_2, NO_2, and O_3 concentrations [21,22].

For Delhi, AQI levels typically fall within the 160–350 range, indicating moderate to unhealthy conditions for sensitive groups, with SO_2 concentrations between 15–40 $\mu g/m^3$ (5–15 ppb), NO_2 between 45–85 $\mu g/m^3$ (25–45 ppb), and O_3 ranging from 60–100 $\mu g/m^3$ (30–50 ppb). In Mumbai, generally better air quality is observed, with AQI values in the 70–140 range and SO_2 levels between 10–25 $\mu g/m^3$ (4–10 ppb), NO_2 between 30–60 $\mu g/m^3$ (16–32 ppb), and O_3 levels of 55–95 $\mu g/m^3$ (28–48 ppb). Kolkata typically experiences AQI levels of 120–240, with SO_2 concentrations of 12–30 $\mu g/m^3$ (5–12 ppb), NO_2 between 35–70 $\mu g/m^3$ (18–35 ppb), and O_3 levels of 50–90 $\mu g/m^3$ (25–45 ppb).

Data from CPCB monitoring networks and analyses from UrbanEmissions.info [23] indicate that air quality has generally worsened over the past decade for these contaminants, particularly in northern cities such as Delhi, where both traffic growth and residential combustion of solid fuels remain dominant emission sources. While Mumbai's coastal location and meteorology moderate extreme pollutant build-up, both Kolkata and Delhi experience seasonal episodes of severe pollution, often linked to biomass burning and regional transport of pollutants. Rising vehicle numbers have also increased NO_2 concentrations year-round and have contributed to an extended "ozone season" in Indian megacities, now often lasting beyond the traditional April–June pre-monsoon period.

The Air Quality Index (AQI) in India follows a scale of 0–500, with values above 200 categorized as "poor" to "severe". Table 4 summarizes recent AQI-based contaminant concentrations in Delhi, Mumbai, and Kolkata from 2022–2023 (with additional data for Ghaziabad and Noida). Values are approximate based on CPCB 2022–2023 and peer-reviewed studies.

3.1 Example: AQI for Delhi

Based on reported Air Quality Index (AQI) and air pollutant monitoring data provided by the Central Pollution Control Board (CPCB) [24] and the System of Air Quality and Weather Forecasting and Research (SAFAR) [25] over the last two decades, the characteristics of air quality, prominent pollutants, and annual variation of average concentrations of SO_2 and NO_2 in Delhi have been analyzed. The average AQI of Delhi in recent years has been consistently among the highest of major Indian metropolitan areas, often exceeding levels observed in cities such as Mumbai, Bengaluru, Hyderabad, and Chennai, and comparable to nearby Ghaziabad and Noida [26]. Concentrations of air pollutants routinely exceed both India's National Ambient Air Quality Standards (NAAQS) and the more stringent World Health Organization (WHO) air quality guidelines [13]). $PM_{2.5}$ and NO_2 pollution are particularly severe, highlighting the growing influence of emissions from vehicular traffic, industrial activity, biomass burning, and regional dust transport [27].

North India, including the National Capital Region (NCR), experiences severe winter spikes in $PM_{2.5}$ and PM_{10} levels due to temperature inversion, stagnant atmospheric conditions, and large-scale crop residue burning in surrounding agricultural states. These seasonal factors are compounded by high vehicular emissions, industrial activity, and open waste burning, which accumulate under low wind speeds. During summer, strong sunlight and high temperatures drive photochemical reactions, resulting in elevated ozone (O_3) concentrations, while dust storms from arid regions add to coarse particulate matter levels, impacting both the NCR and the wider northern region.

Table 4. AQI data expressed as contaminant concentrations for Delhi, Mumbai, and Kolkata (with additional data for Ghaziabad and Noida)

Delhi AQI Data			
Contaminants of Concern	AQI Range Moderate/Unhealthy for Sensitive Groups	Contaminant Concentration Range ($\mu g/m^3$)	(ppb)
SO_2	160 – 350	15 – 40	5 – 15
NO_2		45 – 85	25 – 45
O_3		60 – 100	30 – 50
$PM_{2.5}$		180 – 350	[not available]
Mumbai AQI Data			
Contaminants of Concern	AQI Range Moderate	Contaminant Concentration Range ($\mu g/m^3$)	(ppb)
SO_2	70 – 140	10 – 25	10 – 25
NO_2		30 – 60	30 – 60
O_3		55 – 95	55 – 95
$PM_{2.5}$		60 – 120	[not available]
Kolkata AQI Data			
Contaminants of Concern	AQI Range Moderate/Unhealthy	Contaminant Concentration Range ($\mu g/m^3$)	(ppb)
SO_2	120 – 240	12 – 30	5 – 12
NO_2		35 – 70	18 – 35
O_3		50 – 90	25 – 45
$PM_{2.5}$		100 – 200	[not available]
Ghaziabad AQI Data			
Contaminants of Concern	AQI Range Unhealthy	Contaminant Concentration Range ($\mu g/m^3$)	(ppb)
SO_2	180 – 360	12 – 35	5 – 12
NO_2		40 – 80	22 – 40
O_3		55 – 90	28 – 45
$PM_{2.5}$		190 – 360	[not available]
Noida AQI Data			
Contaminants of Concern	AQI Range Moderate/Unhealthy	Contaminant Concentration Range ($\mu g/m^3$)	(ppb)
SO_2	150 – 320	10 – 30	4 – 10
NO_2		38 – 70	20 – 38
O_3		50 – 85	25 – 42
$PM_{2.5}$		160 – 320	[not available]

4. Health Effects of India's Air Pollution

Ambient air pollution in India has been shown to significantly increase both morbidity and mortality in the general population [26]. The development of respiratory diseases from direct inhalation exposure to air pollutants is well documented as a major contributor to adverse health outcomes [28]. There is strong evidence linking ambient air pollution exposure with increased risk of mortality from cardiovascular and cerebrovascular diseases [29].

Long-term exposure to $PM_{2.5}$, O_3, and NO_2 is associated with increased risks of respiratory infections, cardiovascular diseases, and cancer. The Indian Council of Medical Research (ICMR) reported in 2020 that air pollution was the second leading risk factor for disease burden in India [30].

Indoor air pollution remains a concern, particularly in households using solid fuels. Formaldehyde and benzene concentrations in indoor environments often exceed safe thresholds due to poor ventilation and off-gassing from building materials.

Indoor pollutants, such as formaldehyde, benzene, and total volatile organic compounds (TVOCs), also pose health risks when concentrations exceed established safe thresholds (Table 5) [31]. Overall, air quality in India remains unsatisfactory in many regions, with pollution-related disease burden ranking among the highest globally [26].

Studies by Indian academic and medical institutions report:

- Short-term exposure to high pollutant levels is linked to increased hospital admissions for cardiovascular and respiratory diseases, particularly among older adults [32].
- The relative risk of respiratory disease admissions increases with exposure to SO_2 and O_3, while cardiovascular admissions are strongly correlated with SO_2, $PM_{2.5}$, and PM_{10} [33].
- Positive correlations have been observed among NO_2, O_3, $PM_{2.5}$, and PM_{10} levels, with winter months showing a particularly strong association between O_3 and hospital admissions [34].
- In most studies, residents aged 65 and over were found to be the most vulnerable population group [28].

Table 5. Common air pollutants, properties, health effects, and sources.

Pollutant	Properties	Health Effects	Sources
Formaldehyde	Colorless irritant gas that can be absorbed by the respiratory tract.	- Tearing, throat discomfort, nausea at low concentration and death at high concentration - Cancers of the nasal cavity, oral cavity, throat, skin and alimentary canal could be caused by long-term exposure to formaldehyde	Particle board, plywood, furniture, formaldehyde containing decoration materials, combustion products
Benzene	Colorless transparent oily liquid, volatile, combustible	- Inhibition of human hematopoietic function - May cause anemia, infection, bleeding under the skin; dysfunction of brain, liver, kidney	Motor vehicle exhaust, paint, liquid detergents, wood stain, plastic pipes
Total volatile organic compounds (TVOCs)	Compounds with boiling range between 50~260°C	- Irritation and discomfort of mucous membranes, eyes, at concentrations of 3.0~25 mg/m^3 - Possible neurotoxic effects at concentrations >25 mg/m^3.	Quick-drying paint, cleaning compounds, coatings, adhesives, building materials, detergents, decoration materials, etc.
Fine particles ($PM_{2.5}$)	Particulate matter with mean particle diameters ≤2.5 μm.	- Harmful to the respiratory and cardiovascular systems. - Fine particles can lodge deeply into the lungs. - Can trigger arterial plaque depositions, cause vascular inflammation and atherosclerosis	Motor vehicle exhaust, coal-burning power plants, metallurgy, petroleum, chemical, textile industries

Figure 4 [35] displays average life years lost in North vs. South India using a partition based on the Tropic of Cancer, which cuts through India at 23.5° north of the equator. As per 2020 pollution levels, on the northern side of the partition, Indian citizens lose an average of 6.6 years of life expectancy. In comparison, those who live to the south of the line lose 3.3 years – a staggering difference of 3.3 years in average life expectancy between those who live north and south of the line. While the populations on either side of the Tropic of Cancer partition are similar, some of the poorest states in the country are found on the northern side of the line.

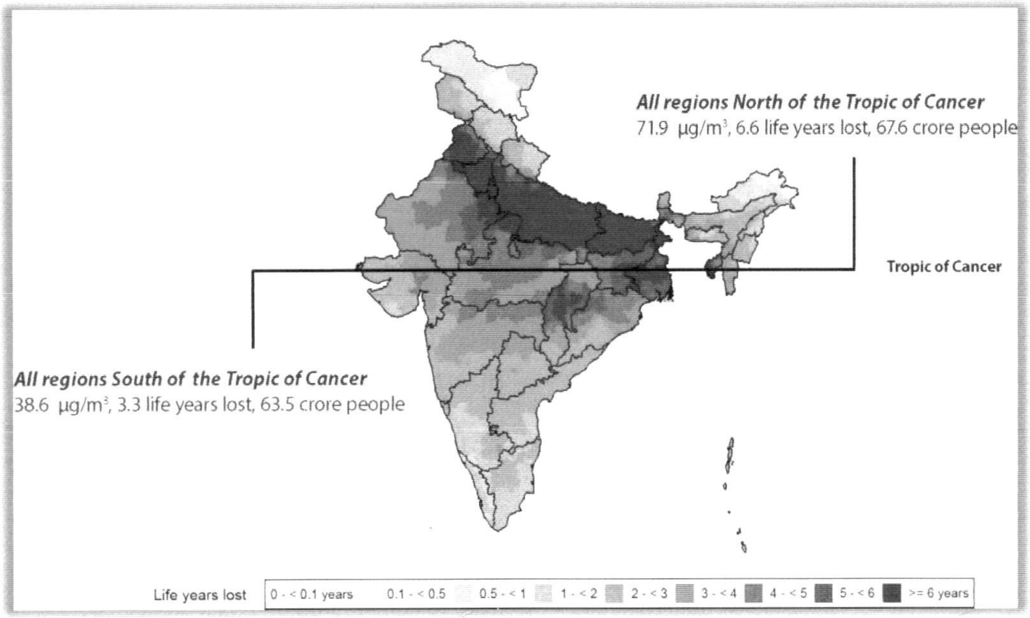

Figure 4. Regional $PM_{2.5}$ exposure and estimated life expectancy loss in India.

Figure 5 shows the regional map estimates of life-expectancy gains across India through permanent $PM_{2.5}$ reduction [36]. In heavily polluted Northern Plains, residents could gain ~8 years of life relative to WHO guideline levels.

5. Enhanced Air Cleaning for Improved IAQ [37]

Removal of irritating and noxious gases and foul odors along with respirable particulate matter ($PM_{2.5}$) are major requirements for any air cleaning systems designed to protect people and assure good IAQ. This includes providing air cleaning technologies for all the indoor environments in which people could be routinely exposed to elevated pollution levels, i.e., inside automobiles, buses, and trains (cabin air), at work, and in the home.

Although the increasing levels of $PM_{2.5}$ present the most serious threat to the health and well-being of those living in India's major urban areas, more emphasis is placed by consumers on the ability of an air cleaning device to remove toxic and irritating chemicals such as HCHO and VOCs. Decisions as to whether or not to install enhanced air cleaning devices for personal use in the home and/or car are being influenced by the growing awareness of the general population of the serious air pollution issues in China, a perceived lack of progress by the central government to adequately address these issues in a timely manner, and being resigned to the fact that ensuring the health of one's family, especially that of children and the elderly, now falls squarely on the individual.

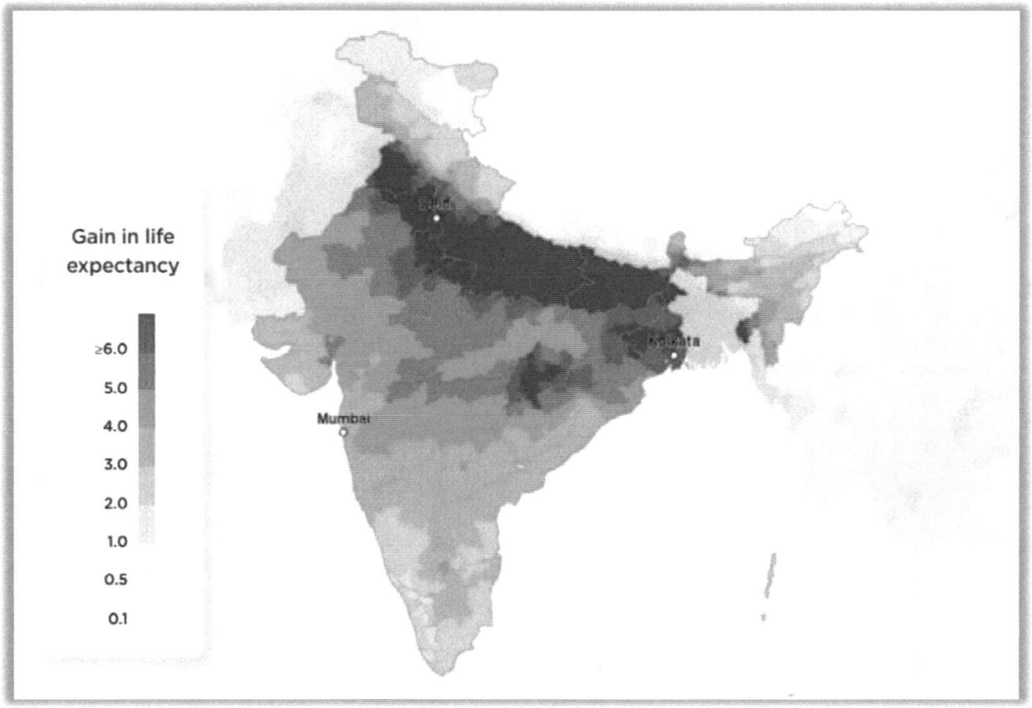

Figure 5. Potential gains in life expectancy through permanently reducing $PM_{2.5}$ from 2018 concentration to the WHO guideline

Among the current air cleaning technologies, there is no single one that can address all primary air pollutants of concern. High-efficiency particulate arrestance (HEPA) filtration for particulate matter and activated carbon for gaseous pollutants are by far the most common technologies in use today; however, most people have a low awareness of the limitations of these two filter types and the benefits of new filter designs and technologies. Ongoing research and development efforts by air cleaner manufacturers has led to the development of enhanced air cleaning devices combining multiple leading-edge air cleaning technologies to effectively remove all common pollutants.

The combination of currently available and new technologies can provide for a more complete solution that addresses the primary ambient air pollutants of concern – these being PM_{10}, $PM_{2.5}$, HCHO, O_3, NO_2, SO_2, and VOCs. However, they cannot be applied as "add on" technologies to existing air cleaning devices. Rather, a new air cleaner design that successfully integrates multiple technologies would prove to be more economical when applied for personal protection, cabin air filters, and general air filtration for the home and workplace.

Indian consumers demand removal of both particulates and gases to mitigate indoor exposure, advanced air purifiers are increasingly being adopted in Indian homes, especially in metros like Delhi and Bengaluru. Many imported air purifiers now include combination filters using HEPA + granular activated caron (GAC) + permanganate-impregnated alumina (PIA).

5.1 Air cleaning for particulate contamination

For the control of particulate matter most manufacturers have incorporated HEPA filtration technology into their air cleaning devices. HEPA filters are generally defined as being able to remove 99.97% of particles ≥0.3 µm in diameter. Given this, HEPA filters would provide for the effective and complete control of the main ambient particulate pollutants, PM_{10} and $PM_{2.5}$.

Filters meeting the HEPA standard have traditionally been used in hospitals to prevent the spread of airborne disease, pharmaceutical and medical device manufacturing to prevent contamination, and the aerospace industry to keep dust off sensitive electronics and optical surfaces. However, more and more they are becoming standard components in filter systems found in commercial aircraft, passenger trains, high-end automobiles, vacuum cleaners, and air purifiers used in the home.

High-efficiency particulate air (HEPA) filters are rated to remove ≥99.97% of particles ≥0.3 µm. Indian-built and imported HEPA units are now widely available for residential and hospital use.

5.1.1 Control of viable particles.

True HEPA filters are very effective at removing viable airborne particles such as fungi, mold spores and bacteria from the air because most of these are larger than 0.3 μm. On the other hand, viruses are so small, some less than 0.03 μm, HEPA filters cannot capture them effectively, although many claim to. These smaller particles will be removed to a lesser efficiency which may not help prevent the spread of airborne disease.

One common misconception is that HEPA filters can kill (or deactivate) harmful microorganisms such as bacteria, fungi and viruses. In fact, most viable particles are simply captured by the filter and many species can live for anywhere from 24-48 hours or more on the filter, and spores can survive indefinitely. To be truly effective against viable particles the HEPA filter must be able to both capture **AND** kill a wide variety of microorganisms.

One major advance in antimicrobial technology has been the development of a particulate filter medium containing silver nanoparticles that acts as an air sanitizer to capture and kill harmful microorganisms. Silver has long been known for its antibacterial properties, but it was only since the development of nanotechnology that allowed creation of materials with embedded silver nanoparticles. A preparation of silver nanoparticles integrated into a HEPA (ULPA[1]) filter medium provides the ability to capture particles as small as 0.3 (0.1) micron and completely inhibit bacterial growth quickly (within a 10-minute contact time). Antimicrobial activity of nanosilver is not limited to bacteria but can control viruses as well. This "capture and kill" technique is effective against bacterial species such as E. coli and S. aureus and viruses such as influenza A and B.

Although the nanosilver does not lose its germicidal properties over time, accumulation of dust, dirt and inactive biomass film can prevent direct contact with the filter medium making it less efficient against targeted organisms. Filter replacement is typically determined by time in service, pressure drop across the filter, or a combination of metrics. HEPA filters combined with silver nanoparticle media have shown promise in reducing bacteria like *E. coli* and *S. aureus* and viruses including influenza. Research at Indian Institute of Technology (IIT) Delhi has evaluated these for hospital use [38].

3.2 Air cleaning for gaseous (chemical) pollution

Activated carbon is the most common adsorbent used in air purifiers to remove chemical contaminants. Prior research by the author [39] has demonstrated the effectiveness of adsorbent media against the chemical compounds in motor vehicle exhaust, one of the primary contributors to India's air pollution problems. The results confirmed that for best results air cleaning systems need to employ at least two different media; activated carbon and permanganate-impregnated alumina.

Among the various delivery systems employed for adsorbent media, the most effective and economical form is 2-3 mm granules or pellets. Granular media provides a higher media weight to volume ratio than commonly available adsorbent-loaded fibers or foams. This in turn provides for high initial and average removal efficiencies along with the longest life for similar size filters.

3.2.1 Granular activated carbon (GAC).

This medium effectively removes VOCs (including benzene, methyl benzene, and dimethylbenzene) and ozone, is only marginally effective against sulfur and nitrogen oxides, and is not effective against formaldehyde. It exhibits a reasonable lifetime and is the least expensive adsorbent medium in wide use.

3.2.2 Permanganate-impregnated alumina (PIA).

While most granular PIA media being used today have available permanganate content of only 4-6%, several media have an 8% permanganate content, and one medium provides a 12% permanganate content which increases performance both in terms of removal efficiency and service life. These media with a higher PIA content are more effective against SO_2, NO_2, and formaldehyde as well as many VOCs and other contaminants of concern for IAQ. Unlike with GAC, pollutants once removed with the PIA medium cannot be released back into the space.

6. New Integrated Air Purifier Designs

Taking advantage of advances in filtration technology, a filter element has been developed that can be used in many different air purifier designs. This combination filter can control all of the primary air pollutants to below the levels specified in India's National Ambient Air Quality Standards (NAAQS), as well as viable particles such as bacteria, mold, and fungi. When employed in an appropriately designed air purifier extremely high single-pass removal efficiencies and a long filter life can be achieved.

Preliminary air purifier testing using a filter element as described here indicates a clean air delivery rate (CADR) for particulate and chemical pollutants approaching 1:1 with respect to the total air volume treated. Efficiencies up to 99.97% for common bacteria and fungus species have been observed when tested for antibacterial activity and efficacy [40].

---

[1] Ultra-low particulate air or ULPA filters; 99.999% efficient for particles ≥100 nanometers (0.1 μm) in diameter.

At design airflows common to commercial and residential air purifiers, the chemical filter options described here exhibit single-pass removal efficiencies ranging from 75-95+% for the contaminants listed above. At the lower air flows typical for personal protective products and cabin air filters, these air purifiers exhibit single-pass removal efficiencies of 90-95% with near 100% efficiency for many of the target contaminants. Using these products as recirculating air cleaners can effectively bring pollutant levels to near non-detectable levels.

Among the current air cleaning technologies, there is no single one can solve all the air pollution problems existing in Indian cities. The two most readily available and economical are HEPA filters and granular adsorbents. With the development of a HEPA filter medium incorporating silver nanoparticles, one can now produce an air purifier that controls both viable particles along with the primary pollutants $PM_{2.5}$ and PM_{10}. Add this level of control to a filter element that employs both GAC and PIA, an air purifier is now available that can also address all the primary chemical pollutants of concern; SO_2, NO_2, O_3, HCHO, and VOCs. Products incorporating this combination of filter media can be applied in masks for personal protection, for cabin air filtration, and in air purifiers for residential and commercial use.

Modern air cleaning units developed for Indian conditions integrate HEPA filtration, GAC, and PIA in a single cartridge system. Tests at the National Environmental Engineering Research Institute [41] indicate CADR values nearing 1:1 (CADR:CFM), and single-pass efficiencies >90% for most common indoor pollutants. Such integrated devices offer a promising solution for improving IAQ across Indian urban and semi-urban settings, especially during peak pollution months from October to February.

7. Conclusion

India's severe air pollution problem requires both systemic outdoor control measures and individual indoor protection strategies. Enhanced air cleaning technologies – especially those that combine HEPA and chemisorbent media – can help mitigate health risks indoors. As public awareness grows, adoption of integrated air purification systems is expected to increase rapidly in Indian cities.

HEPA filters effectively remove $PM_{2.5}$ and PM_{10}, while activated carbon is effective for VOCs, benzene, and ozone, and PIA offers better removal for SO_2, NO_2, and formaldehyde. Filters with higher permanganate content (≥12%) have been tested with success in labs and airports [42].

References

[1] IQAir. World Air Quality Report 2025 – Most Polluted Cities. IQAir, 2025. https://www.iqair.com/world-most-polluted-cities (accessed August 5, 2025).

[2] The Guardian. "The air is killing us: Why Delhi's pollution problem runs deeper than smog season." The Guardian, Nov 22, 2024. https://www.theguardian.com/world/2024/nov/22/the-air-is-killing-us-why-delhi-india-pollution-problem-runs-deeper-than-smog-season (accessed August 5, 2025).

[3] Times of India. "In Delhi, you don't have to smoke to get lung cancer." Aug 2025. https://timesofindia.indiatimes.com/city/delhi/in-delhi-you-dont-have-to-smoke-to-get-lung-cancer/articleshow/123027767.cms (accessed August 5, 2025).

[4] Tracing the Hazy Air 2025: Progress Report on National Clean Air Programme (NCAP). https://energyandcleanair.org/publication/tracing-the-hazy-air-2025-progress-report-on-national-clean-air-programme-ncap/ (accessed August 5, 2025).

[5] Centre for Science and Environment (CSE). State of India's Environment 2023. CSE, New Delhi, 2023.

[6] Central Pollution Control Board (CPCB). National Ambient Air Quality Standards (NAAQS), 2009. https://cpcb.nic.in/uploads/National_Ambient_Air_Quality_Standards.pdf (accessed August 5, 2025).

[7] CPCB. National Ambient Air Quality Standards (NAAQS) Update 2009. https://cpcb.nic.in/ (accessed August 5, 2025).

[8] CPCB. Air Quality Status 2023. https://cpcb.nic.in/about-namp/ (accessed August 5, 2025).

[9] IQAir. World Air Quality Report 2023. IQAir, 2023. https://www.iqair.com/world-most-polluted-cities (accessed August 5, 2025).

[10] Central Pollution Control Board. National Air Quality Monitoring Programme (NAMP). https://cpcb.nic.in/about-namp/ (accessed August 5, 2025).

[11] CSPS India. Air Quality Monitoring Measures in India. https://cspsindia.org/air-quality-monitoring-measures-in-india (accessed August 5, 2025).

[12] CPCB Portal. https://cpcb.nic.in/index.php (accessed August 5, 2025).

[13] World Health Organization (WHO). WHO Global Air Quality Guidelines. Geneva, WHO, 2021.

[14] Guttikunda, S.K., Nishadh, K.A. Urban Emissions in India: Trends and Insights. UrbanEmissions.info, 2022.

[15] Chakraborty, A., Basu, S. Annual PM2.5 Concentration Distribution by District, India 2010. UrbanEmissions.info, 2021.

[16] Ministry of Environment, Forest and Climate Change (MoEFCC). https://moef.gov.in/ (accessed August 5, 2025).

[17] CPCB AQI Portal. https://app.cpcbccr.com/AQI_India/ (accessed August 5, 2025).

[18] CPCB. Air Quality Trends 2014. https://cpcb.nic.in/ (accessed August 5, 2025).
[19] Central Pollution Control Board (CPCB). National Air Quality Monitoring Programme (NAMP) & AQI Data Portal. https://app.cpcbccr.com/AQI_India/ (accessed August 5, 2025).
[20] IQAir. World Air Quality Report 2025 – Most Polluted Cities. IQAir, 2025. https://www.iqair.com/world-most-polluted-cities (accessed August 5, 2025).
[21] Guttikunda, S.K., Nishadh, K.A. Urban Air Quality in India: Trends in $PM_{2.5}$ and Gaseous Pollutants. UrbanEmissions.info, 2022.
[22] Chakraborty, A., Basu, S. Annual $PM_{2.5}$ Concentration Distribution by District, India 2010. UrbanEmissions.info, 2021.
[23] UrbanEmissions.info. India Air Pollution Data and Modeling (1998–2020). https://urbanemissions.info/ (accessed August 5, 2025).
[24] CPCB. Ambient AQ Data - Delhi NCR. https://cpcb.nic.in/ambient-aq-data-delhi-ncr/ (accessed August 5, 2025).
[25] SAFAR-India: System of Air Quality and Weather Forecasting and Research. https://safar.tropmet.res.in/index.php?menu_id=1 (accessed August 5, 2025).
[26] Health Effects Institute. 2024. State of Global Air 2024. Special Report. https://www.stateofglobalair.org/resources/report/state-global-air-report-2024 (accessed August 5, 2025).
[27] UrbanEmissions.info. 2022. Update. https://urbanemissions.info/ (accessed August 5, 2025).
[28] Balakrishnan, K., et al. 2019. The Impact of Air Pollution on Deaths, Disease Burden, and Life Expectancy across the States of India: The Global Burden of Disease Study 2017. *The Lancet Planetary Health*, 3, E26-E39.
[29] GBD India Compare. Global Burden of Disease Study – India Dashboard 2019. https://vizhub.healthdata.org/gbd-compare/ (accessed August 5, 2025).
[30] Indian Council of Medical Research (ICMR). India State-Level Disease Burden Initiative: Air Pollution and Health. ICMR, 2017.
[31] Environmental Protection Department. 2006. Air Study Reports. April 28. www.epd.gov.hk/epd/english/environmentinhk/air/studyrpts/effect_ambient_ap.html/ (accessed August 5, 2025).
[32] Joshi, S., et al. Hospital Admissions and Air Pollution in India. Environmental Research, 2020.
[33] Kumar, R., et al. Air Pollution and Respiratory Risk in India. Science of the Total Environment, 2018.
[34] Guttikunda, S.K., et al. Source Contributions to $PM_{2.5}$ in India. Environmental Science & Technology, 2019.
[35] EPIC India. Air Quality Insights Portal. https://epic.uchicago.in/ (accessed August 5, 2025).
[36] Energy Policy Institute at the University of Chicago (EPIC). Air Quality Life Index (AQLI) – India 2023. https://aqli.epic.uchicago.edu/ (accessed August 5, 2025).
[37] Muller, C.O. and Yu, H. 2015. Ambient Air Quality in China: The Impact of Particulate and Gaseous Pollutants on IAQ. Procedia Engineering, 121: 582-589, PROENG26189, PII: S1877-7058(15)02765-4, doi: http://dx.doi.org/10.1016/j.proeng.2015.08.1037 (accessed August 5, 2025).
[38] Gupta, S., et al. Impact of Air Pollutants on Human Health in India. *Atmospheric Environment*, 2021.
[39] Muller, C. Proceedings of ISHVAC 2007. 2007. Muller, C. 2009. The Control of Ozone and Nitrogen Oxides from Automobile Exhaust for Improved IAQ, In: *Proceedings of ISHVAC 2007*, the 5th International Symposium on Heating, Ventilating and Air Conditioning, September 7-8, 2007, Tsinghua University, Beijing, China.
[40] JIS. 2008. JIS Standard L 1902:2008, Testing for Antibacterial Activity and Efficacy on Textile Products, Tokyo. Japanese Standards Association.
[41] National Environmental Engineering Research Institute (NEERI). Air Quality Monitoring Studies. NEERI, 2023.
[42] Deshmukh, R. NEERI Technical Report: Air Purifier Performance Testing, 2022.

INVESTIGATION OF THE EFFECT OF ACOUSTIC BARRIERS ON DAYLIGHT IN A CLASSROOM

Asvika KS[1,*] , Subbaiyan G [1,2]

[1] National Institute of Technology, Tiruchirappalli, India
[2] National Institute of Technology, Tiruchirappalli, India
[*] 401121001@nitt.edu

SUMMARY

Traffic noise is a significant sound pollutant in urban environments. Previous studies have identified that schools near the roadside are severely impacted by Traffic noise, affecting the teaching-learning process. An educational institution in an urban area adjacent to a busy road is considered for the study. Simulations were carried out for the indoor noise level of an empty classroom, with varying horizontal louvers angles and WWR for different floors. A layer of sound-absorptive material was provided on the bottom side of the louvers, reducing the noise ingress that made most cases comply with IS 8827 for the background noise level threshold for educational institutions. The upward tilt of louvers had a lower SPL (A) than the case without shade. The daylight of the room drastically reduces as the angle of the louvers goes higher. It was noticed that an angle of -15° to 60° complies with the LEED standard of daylighting.

KEYWORDS

Daylighting; Educational Institutions; Traffic Noise; Horizontal Louvers; Simulation

1 INTRODUCTION

In urban areas, traffic noise is considered a major sound pollutant, and it is known that both long-term and short-term exposure to noise has a strong impact on human health and behaviour (Lin Fritschi & Rokho Kim, 2011). Schools that are very close to the roadside are intensely affected by high-pitch noise, significantly affecting the teaching-learning process(El-Sallamy et al., 2018). If installing an exterior noise barrier is not possible or ineffective, the design of the building façade becomes crucial. It is easy to obtain good performances in terms of sound insulation of the opaque components of a building facade, whereas the windows represent the weakest element in the facade's sound insulation.

There is a direct relationship between the area of window openings and their characteristic level of acoustic insulation; larger openings provide poorer acoustic protection (Waters-Fuller et al., 2006). Placement of acoustic shades, in turn, affects the daylight that enters through the window. Therefore, it requires a solution that provides adequate daylighting indoors and better noise reduction, i.e., reduces outdoor noise from entering the classroom. The study's main objectives are to identify the effect of window size and shading system on daylight and noise level, and to determine an appropriate solution that can provide adequate daylight and maximize noise reduction in the classroom.

2 METHODS

2.1 Study Location and Description of Classroom

The study focuses on the perception of visual and acoustic comfort in classrooms. This study involves a preliminary simulation of a classroom in typical conditions. The prototype model was made using a classroom of a school building located in an urban area adjacent to busy roads in Chennai (13.08°N, 80.27°E), India. Chennai falls under warm and humid climate zone, as per NBC (*National Building Code*, 2016) . The building is adjacent to two busy roads, and at least one façade of the classrooms was exposed to traffic noise. A typical classroom is square in shape, measuring 8m x 8m, with an area of 64 m^2, and the height of the ceiling is 3.9m. The classrooms followed a typical layout; hence, the same model was used for further simulations by considering the Sound Pressure Level of traffic noise as 70dB (*National Building Code*, 2016).

2.2 Simulation and Questionnaire Survey

A prototype model of typical dimensions and other material specifications was made, and simulations were carried out to check how the tilt of louvers, WWR, and height from the ground affect the Acoustic and Daylight in the classroom. Grasshopper was used for simulating different louver settings and their impact on daylighting and acoustics in the classroom. Compliance with the LEED Standard for Daylighting (LEED, 2014) (i.e. 40% of the floor area should have illuminance level of 100 lux to 2000 lux for at least 90% of potential daylit time) was done for the classroom for varying louver angles and fixed WWR of 0.2, which is the existing WWR of the classroom. An Annual Glare Analysis was done for the classroom to identify the positions of glare within the classroom.

Acoustic Parameters like Leq, Clarity, and RT were calculated using Pachyderm. Noise Reduction was calculated using equation 1.

$$NR = TL - \log_{10}\left(\frac{S_W}{\alpha}\right) \qquad \text{Eq. (1)}$$

Where TL is the Transmission Loss (dB) of the walls, S_w is the surface area (m^2) of the walls between two rooms, and α is the total absorption of the adjacent room (m^2).

The Daylight Metrics like UDI2000, DGP, DGI, Illuminance Level, etc., were simulated using Honeybee, Grasshopper and Climate Studio. The cross-modal effect of the external shading System on lighting and noise reduction was analysed using Excel and SPSS. Analysis also balanced the external shading system on the lighting and acoustic Parameters specified.

A Questionnaire survey was conducted to evaluate the visual perception of the classroom for the finalized louver angle. The study involved fifty participants from various age groups, who were asked to evaluate their perception of daylight in the space. The questionnaire consisted of various simulated images of classroom conditions, and questions to assess the perception of the brightness, uniformity of distribution, glare and visual comfort. The respondents rated parameters such as brightness, uniformity, glare, and overall visual comfort based on simulations of the classroom with different louver configurations, two viewpoints, and two different months. A Wilcoxon signed-rank test was performed to assess the impact of an acoustic shutter on daylight conditions in a classroom.

3 RESULTS AND DISCUSSIONS

3.1 Daylight Results

Shading systems with horizontal louvers of different inclinations were considered for this study (Figure 1). The surface reflectance of the wall, ceiling, floor, and louver considered for simulations is 0.6, 0.8, 0.3, and 0.5, respectively. The annual sun exposure (in hours) of the room drastically reduces as the angle of the louver goes higher, as seen in Figure 2. There is little to no variation of daylight when compared floor-wise, but that could change in an actual case study where the surroundings will be considered.

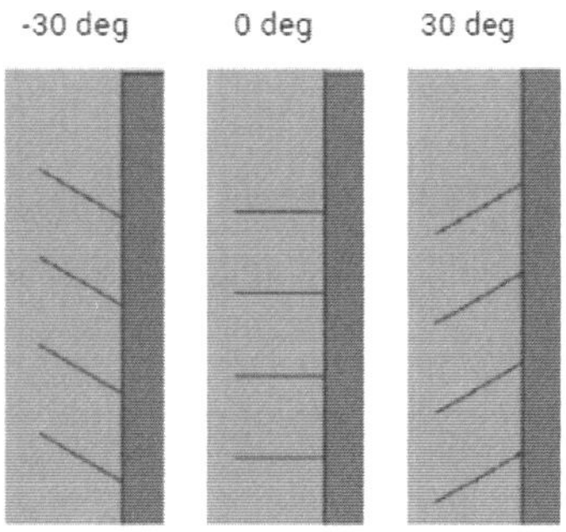

Figure 1. Varying Tilt of Louvers

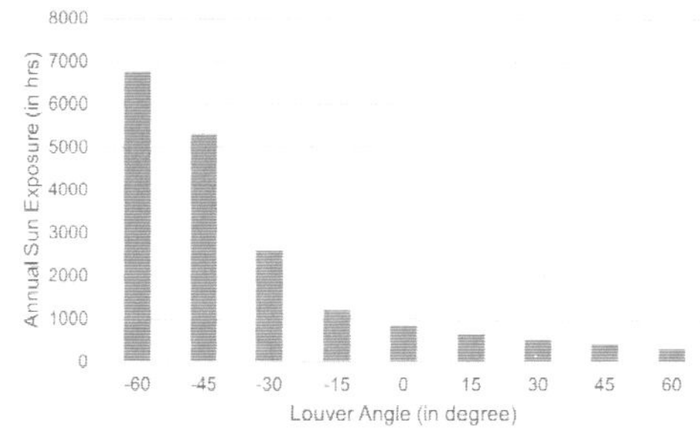

Figure 2. Annual sun exposure for different tilt angles

3.1.1 LEED Compliance

According to LEED Standards, no more than 10% of the regularly occupied space should receive more than 1000 lux of direct sunlight for over 250 hours per year. It is observed that from angle -15° to 60° (Table 1) complies with the LEED standard of daylighting (LEED, 2014) (<10% of floor area). So, for further simulations, the louver angle of -30 ° to 60° was considered.

Table 1. LEED Compliance Analysis

Tilt	-60°	-45°	-30°	-15°	0°	15°	30°	45°	60°
Annual Sun Hour Exposure(in hrs)	6754	5314.7	2615.5	1222.7	860.0	643.7	530.0	432.5	318.5

% Floor area more than 250 hours	17%	15%	3%	0%	0%	0%	0%	0%	0%

3.1.2 Glare Analysis

To analyse the glare in the classroom, a simulation and calculation of daylight glare probability were performed using Climate Studio and Grasshopper. DGP is the most robust metric that generates the most plausible results under the investigated scenes and daylighting conditions. DGP responds predictably to most daylight situations. The recommended DGP must be less than 0.35 for Imperceptible Glare. Table 2 shows the result of glare analysis done for louver angles -30 to 60. It can be noticed that Daylight Glare Probability in all the cases is <0.35, which is imperceptible glare. This might be due to the use of louvers or due to unilateral lighting.

Table 2. Daylight Glare Probability (DGP) Analysis

WWR / Tilt	-30°	-15°	0°	15°	30°	45°	60°
0.2	0.249	0.249	0.251	0.251	0.244	0.226	0.118
0.3	0.249	0.252	0.256	0.257	0.245	0.232	0.173
0.4	0.244	0.252	0.259	0.259	0.244	0.229	0.191
0.5	0.247	0.257	0.269	0.268	0.250	0.231	0.193
0.6	0.265	0.273	0.284	0.282	0.261	0.245	0.224

ECBC states that, for educational institutions, more than 40% of the floor area should have an illuminance level of 100 lux to 2000 lux for at least 90% of the potential daylit time. Simulations were done to analyse the illuminance levels of the classroom at the work plane level. Figure 3 shows the illuminance level distribution of the classroom in summer and winter. UDI analysis was done for various louver tilts to check for ECBC compliance (*BEE_ECBC, 2017*) . Table 3 shows the floor area percentage that has an illuminance level of 100 to 2000 lux for at least 90% of potential daylit time. The louver angles, -30°, -15°, 0°, and 15° comply with ECBC requirements for UDI. Hence louver angles, -30°, -15°, 0°, and 15° were used for further analysis.

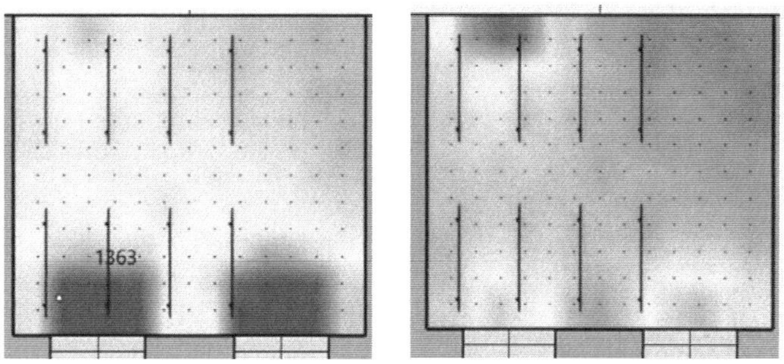

Figure 3. Illuminance Distribution of the Classroom in Summer and Winter

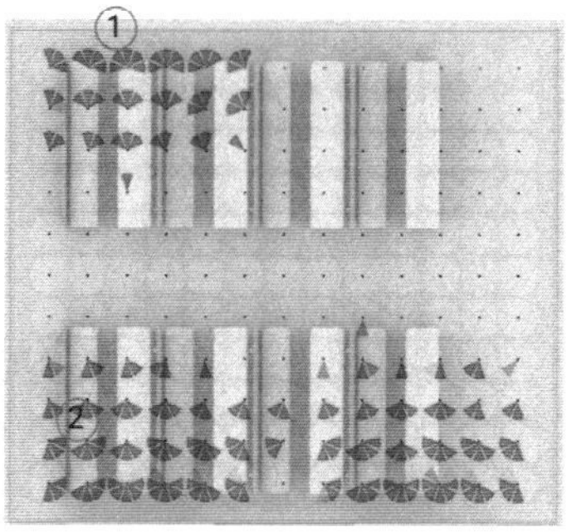

Figure 4. Annual Glare Analysis

Table 3. ECBC compliance for daylight

Angle of Louvers	% of floor area meeting UDI requirement	ECBC Compliance
-30°	66.43	Yes
-15°	81.12	Yes
0°	94.41	Yes
15°	67.13	Yes
30°	23.78	No
45°	10.49	No
60°	4.90	No

An annual glare analysis was done for the classroom to identify the positions of glare that could be used for simulating views. Figure 4 shows that views 1 & 2 were determined to have high glare in the cone of vision of the board. Further glare analysis of the identified windows was done individually. It was identified from Figure 5 that View 1 (Northern windows) has glare during summer, whereas View 2 (Southern window) has glare during winter.

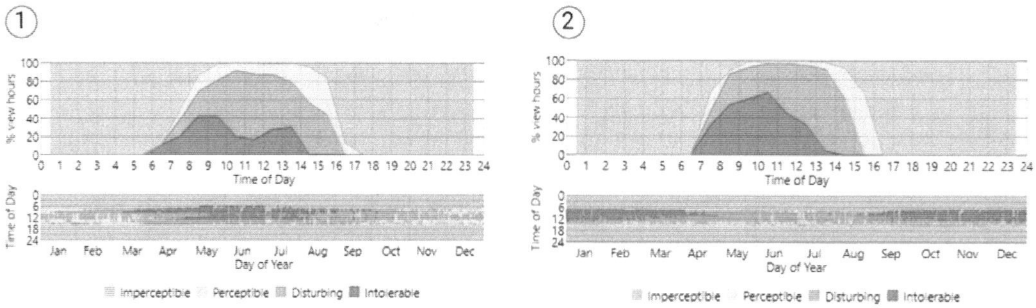

Figure 5. Annual Glare Analysis (1) Northern Window (2) Southern Window

3.2 Acoustics Results

A layer of melamine foam on the lower side of the louvers was used for all acoustic simulations. The absorption coefficients for walls, ceiling, floor, and absorptive layer are 0.035, 0.03, 0.03, and 0.85, respectively, and are used in simulations. Sound pressure level A-weighted was calculated for varying louver tilt with fixed louver dimensions and count. The upward tilt had a lower SPL (A) compared to the downward tilt and without shade. It could be seen that louvers have significantly reduced the noise penetration from outside.

3.2.1 Indian Standard Compliance

With a Sound Pressure Level of outdoor noise as 70 dB (*National Building Code*, 2016), the indoor noise level of an empty classroom is simulated, with varying Horizontal louver angle and WWR for different floors. It can be noticed that most of the cases comply with IS 8827 for the background noise level threshold for educational institutions. 35 dBA criterion recommended by the American National Standards Institute (*ANSI, 2005*) for educational settings. Maximum acceptable noise levels in classrooms and other teaching spaces due to external sources should be 40 dB(A) (Bureau of Indian Standards, 1978) .

3.3 QUESTIONNAIRE SURVEY

A questionnaire survey was conducted with 50 participants, who were asked questions about perception of brightness (Q1), uniformity (Q2), Glare(Q3), and overall visual comfort (Q4) of the space. The images were taken from 2 selected views (L & R), in two seasons (J & D), and the participants were asked to evaluate them. The brightness significantly reduces in all 4 cases when a proper shading device is used. Also, glare is significantly reduced when proper shading is used. The visual comfort has increased after the use of shading devices, and also the same for uniformity of daylight distribution.

4 CONCLUSIONS

The study focused on the perception of visual and acoustic comfort in classrooms. A prototype model of typical dimensions and other material specifications was made, and simulations were carried out to check how the tilt of louvers, WWR, and height from the ground affect the acoustics and daylight in the classroom. The daylight level of the room drastically reduces as the angle of the louver goes higher. Louver tilt angle -15° to 60° complies with the LEED standard of daylighting and has imperceptible glare in DGP analysis. The louver angles, -30°, -15°, 0°, and 15° comply with ECBC requirements for UDI.

A layer of melamine foam on traditional horizontal louvers was observed to have significantly reduced the noise penetration from outside into the classroom. It was identified that the louver angles, -30°, -15°, 0°, and 15° comply with IS 8827. The upward tilt had a lower SPL (A) compared to the downward tilt and without shade. It was concluded that -15° and 0° tilt of louvers provide optimum daylight and noise reduction in the case of the given scenario and satisfy compliance with standards of daylighting and acoustics.

Annual glare analysis of the classroom shows that Views 1 & 2 have high glare in the cone of vision of the board, also View 1 (Northern windows) has glare during summer, whereas View 2 (Southern window) has glare during winter. These simulated images were used in a questionnaire survey where participants were asked questions on perception of brightness (Q1), uniformity (Q2), Glare(Q3), and overall visual comfort (Q4) of the views in different seasons. The impact of the shading device was also analysed in the survey. Brightness and glare were reduced, whereas uniformity and overall visual comfort were seen to increase with the use of a shading device in the louver tilt angle of -15°.

5 REFERENCES

American National Standards Institute (2005).

Bureau of Energy Efficiency : Energy Conservation Building Code (2017).

Bureau of Indian Standards (with Public.Resource.Org). (1978). *IS 8827: Recommendations for basic requirements of school buildings*. http://archive.org/details/gov.in.is.8827.1978

El-Sallamy, W., Osman, T., & Abbas, S. (2018). Acoustic consultant – An under-development novel indoors and outdoors noise assessment software. *Ain Shams Engineering Journal*, *9*. https://doi.org/10.1016/j.asej.2017.12.002

LEED, U. (2014). *Green Building 101: What is indoor environmental quality? | U.S. Green Building Council*. https://www.usgbc.org/articles/green-building-101-what-indoor-environmental-quality

Lin Fritschi, A. L. B., & Rokho Kim, D. S. (Eds.). (2011). *Burden of disease from environmental noise: Quantification of healthy life years lost in Europe*. World Health Organization, Regional Office for Europe.

National Building Code. (2016). Bureau of Indian Standards. https://www.bis.gov.in/standards/technical-department/national-building-code/

Waters-Fuller, T., Lurcock, D., Mackenzie, R., & MacKenzie, R. (2006). *Open/closed window research: Sound insulation through ventilated domestic windows*. https://napier-repository.worktribe.com/output/246785

Impact of Sun Path on Thermal Profile of Indoor Environment in School Classrooms in Malaysia

Yassir Arafat[1], Sheikh Ahmad Zaki[1]*, Noor Alam[1], Naja Aqilah[1], Fitri Yakub[1], Teh Khar San[2], Yam Kah Wei[2].

[1]Department of Mechanical Precision Engineering, Malaysia-Japan International Institute of Technology, Universiti Teknologi Malaysia, Kuala Lumpur 54100, Malaysia
[2]San Miguel Yamamura Woven Products Sdn. Bhd., Lot 9 & 10, Jalan Usaha 4, Kawasan Perindustrian Ayer Keroh, 75450 Ayer Keroh, Melaka

*Email: sheikh.kl@utm.my

ABSTRACT

The thermal performance of building envelopes, particularly roofs, is crucial for indoor thermal comfort in tropical climates like Malaysia. Roofs are the main pathway for solar heat gain, and elevated indoor temperatures can negatively impact student comfort and learning outcomes. The study aims to provide in-situ measurements of the thermal profile across roof assemblies in two secondary schools, capturing temperature variation at multiple layers and quantifying heat transfer into classrooms. Several temperature measuring sensors are placed from 0.5 m above the floor to the rooftop to record daytime data. Understanding vertical temperature gradients is crucial for improving indoor environmental conditions and reducing the amount of heat that is gained in classrooms by improving tactics regarding radiant barriers, roof insulation, and attic ventilation.

Keywords: indoor temperatures, Vertical Thermal Gradient, school classrooms

1. Introduction

The thermal performance of the building envelope, especially roofs, plays a crucial role in ensuring indoor thermal comfort in naturally ventilated classrooms in tropical climates like Malaysia. Roofs serve as the main pathway for solar heat gain, and elevated indoor temperatures may adversely impact on student comfort and learning outcomes. Researchers found a significant relationship between the physical environment and school productivity. Environmental comfort is one of the key elements that has an impact on academic achievement [1]. In recent research, researchers have mentioned that students' performance increases with thermal perception and classroom conditions [2]. More precisely, thermal comfort in a classroom significantly impacts students as they spend more than half of the day in the school. Thermal comfort is defined as a state of mind regarding satisfaction, varied with personal and ambient parameters of the thermal environment [3]. Several researchers have examined the effect of thermal comfort on room temperature. A specific temperature cycle between 25 and 28°C can create thermal comfort by reducing energy consumption [4]. Temperature variation on different scales can be a reason for an unstable thermal environment in the classroom, which increases thermal discomfort. Moreover, the classroom temperature also does not suddenly increase drastically; it is changing over different times of day. Researchers found that the transmission of solar radiation indoors significantly impacts the thermal comfort of individuals near windows [11]. It is imaginable how much the temperature influences the classroom environment without any built-in thermal insulation facilities. Since the thermal performance of a building is highly impacted by the solar absorbance of the roof, it is necessary to consider what type of roof materials are used for the building. The researcher mentioned that between 20% and 90% of radiation is absorbed in the roof surface, and it can reach up to 1 kW/m^2 during clear sky [12]. Metallic roof tile is used as roofing material in those schools. The metallic roof exhibited the highest outdoor temperature when compared to ceramic and concrete roofs [13]. The maximum indoor temperature was recorded on the metallic roof, and the highest heat flux observed among the three types of roofs was also on the metallic roof.

According to Malaysian Meteorological Department (MET) Malaysia, the outdoor temperatures in May 2025 are between 25.4 and 33.4°C, on average, while the weather is tropical, which is mostly humid and hot. The researcher claimed that roof material and colour directly impact the solar heat gains, which affect the indoor thermal comfort conditions of public school buildings in Malaysia [5]. To improve indoor environmental quality and optimize HVAC systems, it is crucial to know the vertical temperature gradients, as they readily impact occupants' comfort.

This study aims to provide in-situ measurements of the thermal profile across roof assemblies in two secondary schools, capturing temperature variation at multiple layers and quantifying heat transfer into classrooms. The vertical temperature distribution in a classroom starts from 0.5 m above the floor to the rooftop. Although according to ASHRAE Standard 55, it is required to start measuring the room temperature from 1.0 m above the ground, in this study, it is considered 0.5 m above the ground floor.

2. Methodology

2.1 Location and climate

The study was conducted in three classrooms, each at two secondary schools, the classroom temperature and vertical thermal gradient were analyzed for SMK Taman Seri Rampai (3°11'31.2"N 101°43'57.9"E) and SMK Padang Tembak (3°10'48.5"N 101°43'24.8"E), Kuala Lumpur, Malaysia. Average toasty temperature feels stuffier due to its location in a hilly equatorial area. Field measurements were carried out from May 15 to July 11, 2025, on the top floor (4th) of both the school buildings, which are situated facing the south direction. The sun's path significantly impacts both schools' indoor temperatures. Solar radiation plays a crucial role in shaping indoor temperatures, primarily by enhancing heat gain through building envelopes, particularly in regions that experience considerable sunlight exposure [6].

Fig.1 (a and b) helps to assume how directly the classroom is heated during the day. Moreover, data from MET Malaysia confirms that the highest temperatures this year were recorded from March to May. The effects of solar incidence on indoor environments and their occupants are substantial, influencing indoor thermal comfort and potentially causing unwanted overheating throughout the year [7]. In Recent years, the mean air temperature and relative humidity in Kuala Lumpur were 27.7°C and 85%, respectively [8].

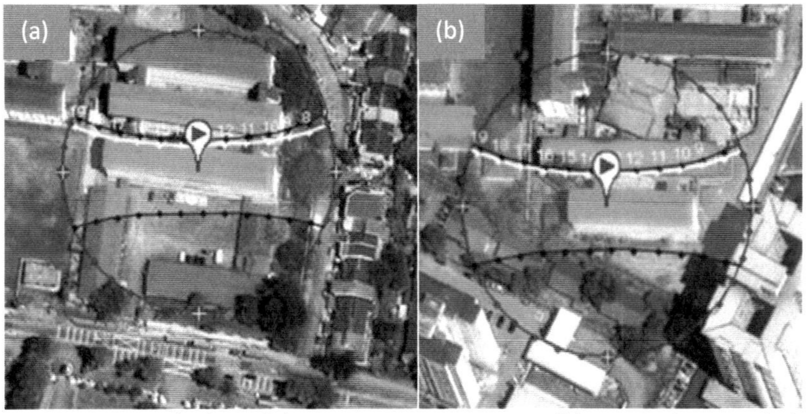

Fig. 1. Sun path above (a) SMK Taman Seri Rampai (b) SMK Padang Tembak.

2.2 Selected schools and classrooms

Large Standard size of classroom in Malaysia, size = Height (H) × Length (L) × Width (W) = 3.60 m × 9.23 m × 7.31 m = 242.89 [9]. The Public Works Department (PWD) decides on the building materials and specifications for public school buildings, which are followed by the contractor, where mostly common building materials are used without considering any thermal insulation properties. The thermal conductivity of common building insulation materials increased approximately linearly with increasing temperature, which is characteristic of materials and varies depending on relative humidity [10].

2.3 Field measurement procedure

For the measurement purpose, a top-floor classroom was selected for both schools, which is in the middle position of the building. Public schools in Malaysia are equipped with louvred windows that were open during measurements to allow natural ventilation. Although the classroom had three ceiling fans in the middle and two wall fans on the left side, those were only used during class hours. The indoor globe and radiant shielded temperature are measured by the HOBO U10 data recorder as shown in Fig.2. Measurement was conducted in three separate time frames, daytime, nighttime, and the whole day, to collect sufficient data.

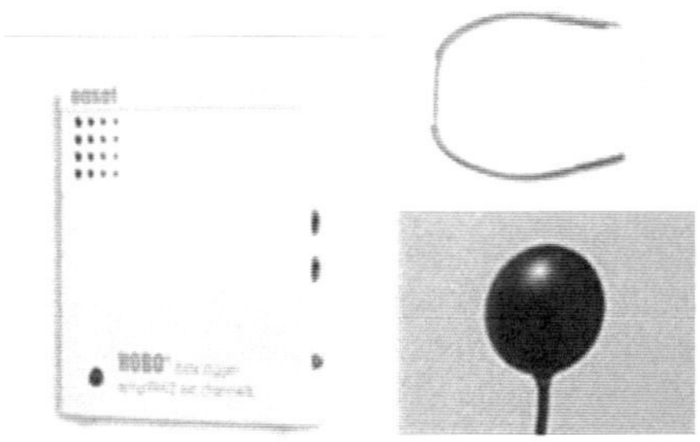

Fig. 2. Hobo indoor temperature recorder U12-013.

The indoor air temperature (T_i) and relative humidity (RH_i) were measured at three heights (0.5 m, 1.5 m, and 2.5 m) to monitor thermal stratification. Moreover, thermocouples are installed in the attic air space, inner ceiling surface, outer roof surface, and inner roof surface to measure temperature variation.

Fig. 3. Indoor temperature measurement.

Besides, to measure outdoor temperature (T_o), two thermocouples were installed outside the classroom. Fig.3 illustrates the position of Hobo for indoor temperature at different heights. Thermocouple is a type of temperature sensor; its low cost and self-powered properties make it preferable for having a stable reference temperature. It consists of two different thermoelements that allow the generation of a potential difference or voltage based on the thermoelectric phenomenon [14]. Heat Flux Transducer: Installed on the ceiling surface to measure heat flow into the classroom. All equipment is set up for data collection in a 5-minute interval, and all thermocouples and heat flux sensors were connected to a Graphtec data logger GL820 Fig.4. HOBO sensors for indoor and globe temperatures are recorded at similar time intervals. The position of sensors above the ceiling is illustrated in Fig.4.

Fig. 4. Data logger GL820

3. Results and Discussion

Fig.5(a). shows that for SMK Sri Rampai, room temperature was almost the same for three different heights, which start below 30°C, and except for a few hours of daytime, any drastic change is not visible. Rooftop temperature shows fluctuation starting near 8:00, and it peaks above 65°C after 14:00, then a sudden fall in temperature below 30°C is noticed around 15:00. Although a temperature rise happens at 4:00 pm, it gradually decreases. After 6:00 pm, the temperature from the ceiling to the room decreased, but the room temperature increased. Fig.5(b) SMK Padang Tembak also illustrates similar results with small variations, where roof temperature crossed over 70°C, and above the ceiling, more fluctuation is visible during the daytime. From both graphs, it is common for the rooftop temperature to decrease at night to below 25°C, which is the lowest in the graph. Still, during the daytime, it reaches almost double the average daily temperature in Malaysia.

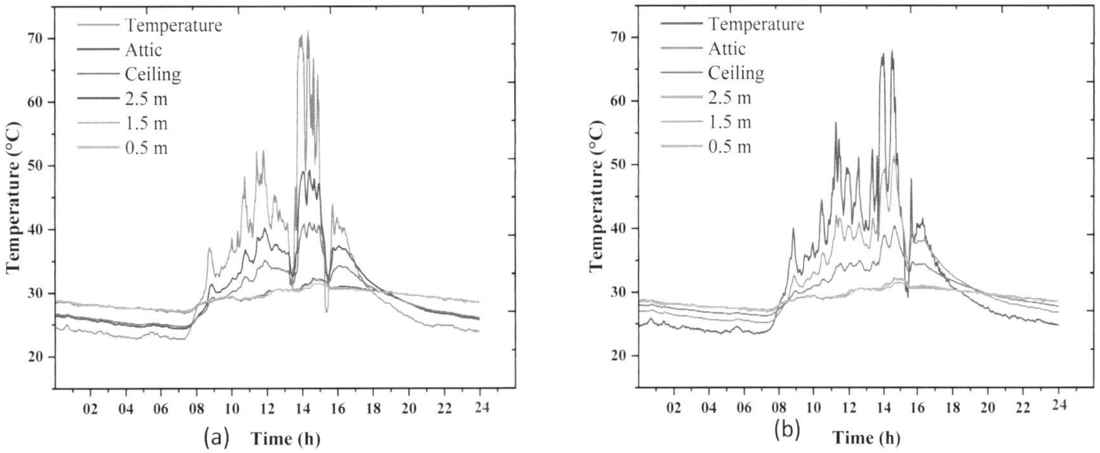

Fig. 5. Hourly average temperature (a) SMK Taman Seri Rampai (b) Roof SMK Padang Tembak

3.1 Comparison of vertical gradient temperatures

In terms of natural ventilation, decreasing solar heat absorption in the roof could contribute to moderate fluctuations in indoor temperature [15]. The vertical gradient of both schools is calculated from the temperature difference between the ceiling and the roof. The vertical gradient (∇T),

$\nabla T = \Delta T / \Delta Z$

ΔT = Temperature difference between two points (°C).

Δz = Vertical distance between the two points (m).

Vertical profiles of the air temperature from floor to rooftop of the classroom are shown in Fig. 6. The present values are the relative temperatures concerning the floor of the classroom, time average of the entire period, including the daytime (7:00-19:00) and nighttime (19:00-7:00). The temperature on the inner surface of the ceiling board exhibited variations between nighttime and daytime conditions. Throughout the daylight hours, the temperature demonstrated a notable rise with increasing elevation, reaching its maximum at the inner surface of the roof tile. The temperature differential between the attic and the inner roof surface was notably substantial, attributed to the heating effects of solar radiation. The roof tile's outer surface temperature was lower than that of the inner surface. This discrepancy is likely attributed to transient evaporation cooling occurring during showers, as well as the heat storage properties of the tile. On the other hand, the temperature above the ceiling during the nighttime exhibited a considerable decrease along the height direction, probably owing to the cooling of the nocturnal radiation. The room air temperature is 6°C and 8°C higher than the outer roof surface temperature during the nighttime for SMK Taman Seri Rampai (SMK-SR) and SMK Padang Tembak (SMK-PT). As a result, SMK PT is much hotter than SMK SR, even in the early morning.

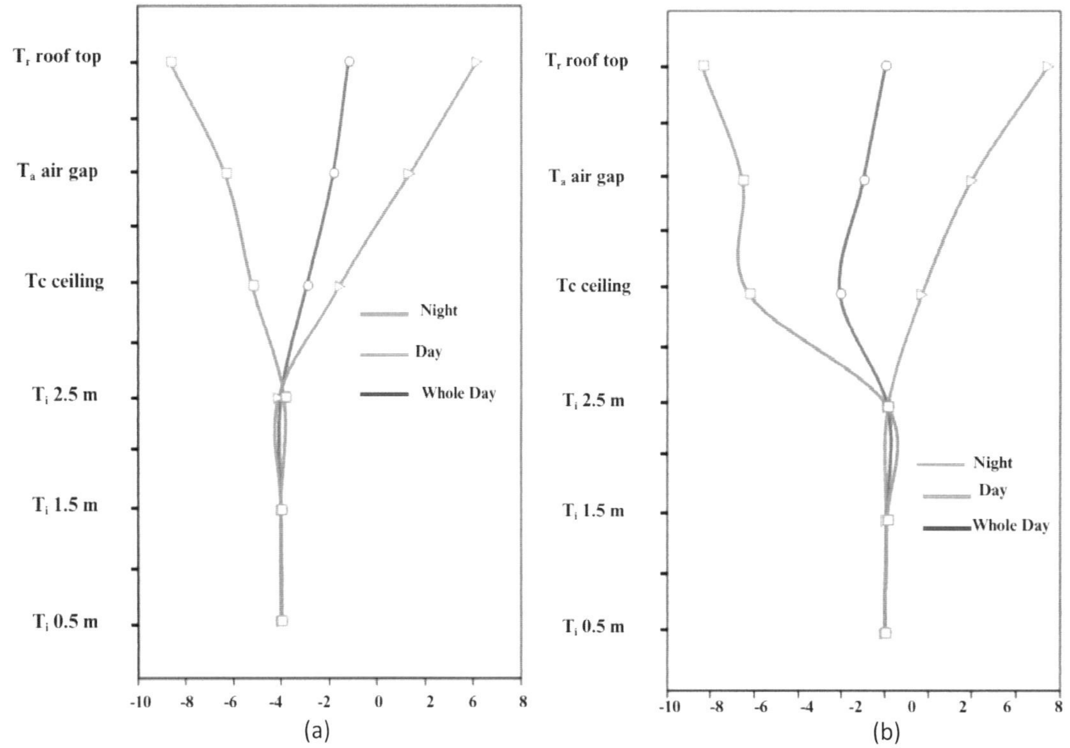

Fig. 6. hourly average temperature (a) SMK Taman Seri Rampai (b) SMK Padang Tembak

3.2 Indoor temperature

Fig.7(a). The graph shows the indoor temperature variations of three classrooms (Room 1, Room 2, and Room 3) in SMK-SR. The temperature fluctuates between 27°C and 36°C in all rooms, with Room 3 (green line) generally experiencing higher temperatures than the other two. All rooms follow a similar periodic pattern, likely influenced by daily factors such as sunlight, air conditioning, or ventilation, with temperatures peaking during the day and dropping at night. Room 3 exhibits the largest fluctuations, reaching 36°C, while Room 2 (orange line) stays lower, ranging from 28°C to 34°C. Room 1 (blue line) falls between the other two, fluctuating between 29°C and 34°C. These findings suggest that while all rooms experience similar temperature cycles, Room 3 tends to be the warmest, indicating that improvements in ventilation or cooling systems might be necessary to enhance comfort in this room. On the other hand, Graph Fig.7(b). illustrates the temperature fluctuations in SMK-PT classrooms. Room 1 (blue line) shows moderate temperature changes, ranging from 28°C to 35°C, with a consistent periodic rise and fall, and slight increases towards the middle of the observed period. Room 2 (orange line) experiences more erratic fluctuations, with temperatures ranging from 29°C to 34°C, showing sharper peaks around mid-June and late June. This suggests varying external conditions or HVAC inconsistencies. Room 3 (green line) displays the widest temperature range, from 27°C to 35°C, with dramatic peaks and dips, especially between mid-June and early July. The temperature variations in all rooms peak during the day, likely due to external factors such as sunlight or internal heat from activities or equipment. Overall, the temperatures in the rooms appear to rise during late June and early July, possibly due to warmer outside weather or changes in the classroom climate.

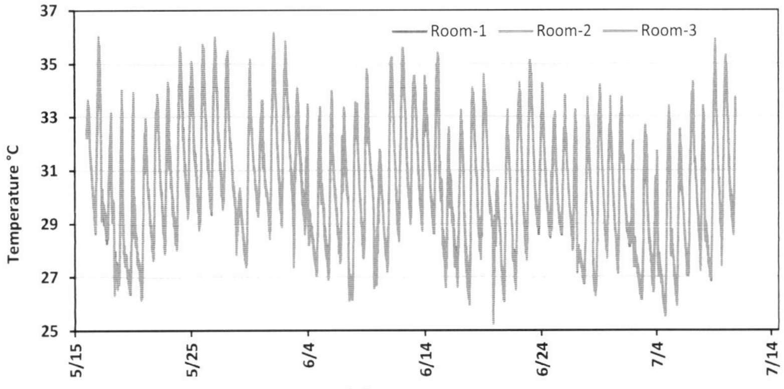

(a)

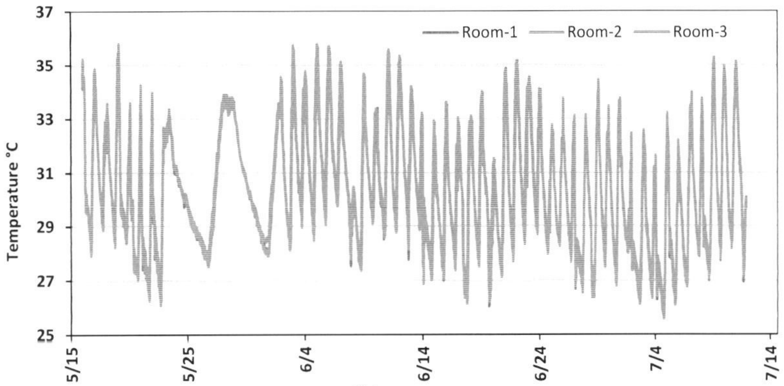

(b)

Fig. 7. Variation of indoor Temperature (a) SMK-SR, (b) SMK-PT

Table 1 presents indoor temperature statistics for school classrooms in Kuala Lumpur, specifically comparing data from SMK Taman Seri Rampai and SMK Padang Tembak. Across the three classrooms at SMK Taman Seri Rampai, maximum temperatures ranged from 35.6°C to 35.8°C, while minimum temperatures varied between 25.6°C and 26°C. The mean temperatures were relatively consistent, ranging from 30.2°C to 30.6°C, with standard deviations (SD) around 2.1°C to 2.2°C, and coefficients of variation (CV) between 7.0% and 7.3%, indicating moderate variation in indoor temperatures. At SMK Padang Tembak, although all entries are labeled "Classroom-1," the readings suggest measurements from different rooms or periods. Maximum temperatures reached as high as 36.2°C, with minimums as low as 25.3°C, and mean temperatures consistently around 30.2°C to 30.2°C. Standard deviations here ranged from 2.2°C to 2.3°C, and CV values were slightly higher, between 7.40% and 7.54%, indicating slightly more temperature variability than SMK Taman Seri Rampai. Overall, both schools experienced similar thermal conditions, with minor differences in temperature stability.

Table. 1 Indoor Temperature Statistics for School Classrooms in Kuala Lumpur.

SMK Taman Seri Rampai					
Room	Max Temp (°C)	Min Temp (°C)	Mean Temp (°C)	*SD* (°C)	*CV* (%)
Classroom-1	35.7	25.6	30.6	2.2	7.3
Classroom-2	35.6	25.6	30.2	2.2	7.3
Classroom-3	35.8	26.0	30.5	2.1	7.0
SMK Padang Tembak					
Classroom-1	35.9	25.5	30.2	2.2	7.4
Classroom-1	36.2	25.4	30.2	2.3	7.5
Classroom-1	36.0	25.3	30.2	2.3	7.5

Note: Max Temp: maximum temperature; Min Temp: minimum temperature; *SD:* standard deviation; *CV*: coefficient of variance

4. Conclusion

This research shows extensive temporal and geographic temperature data gathered from the roof assemblies of classrooms in Malaysian secondary schools. Although the result shows that classroom of SMK Padang Tembak was warmer than SMK Taman Seri Rampai but overall, the data was acquired from the roof assemblies of classrooms suggest that there are significant temperature differences between the metal deck layers of the roof, as well as a delay in the flow of heat into the classrooms, which contributes to an increased level of thermal discomfort. Variations in attic ventilation likely have a substantial impact on the thermal performance of the building. To reduce the amount of heat gained in classrooms, the statistics suggest a need for improved tactics regarding radiant barrier, roof insulation, and attic ventilation. In order to provide a more comprehensive knowledge of seasonal effects and to inform retrofit recommendations targeted at improving energy efficiency and thermal comfort in school buildings, additional analysis will be conducted through increased data collection.

Acknowledgement

The Universiti Teknologi Malaysia financially supported this work through the F4+ High Tech Grant [Vot 00Q82]. This research was funded by a grant from the Ministry of Higher Education of Malaysia (FRGS Grant R.J130000.7824.4X172).

References

[1] Wong, L. T., Mui, K. W., and Tsang, T. W. "An Open Acceptance Model for Indoor Environmental Quality (IEQ)." Build. Environ. 142, (2018): 371–378.

[2] H. Azemati, Z. Aminifar, and S. Pourbagher, "Effective environmental factors on designing productive learning environments," Armanshahr Architecture & Urban Development, 11(22), 1-8, Spring 2018.

[3] C. Z. Y. Chong Zi Yao, M. N. A. Nor Azl, A. Hariri, A. A. Muhamad Damanhuri, and M. S. S. Mustafa, "Preliminary study on student's performance and thermal comfort in classroom," Journal of Advanced Research in Fluid Mechanics and Thermal Sciences, vol. 101, no. 1, pp. 59-72, 2023.
[4] S. K. Sansaniwal, J. Mathur, and S. Mathur, "Review of practices for human thermal comfort in buildings: present and future perspectives," International Journal of Ambient Energy, vol. 43, no. 1, pp. 2097-2123, 2022.
[5] S. Li, X. Jia, C. Peng, Y. Zhu, and B. Cao, "Effects of temperature cycles on human thermal comfort in built environment under summer conditions," Science of The Total Environment, vol. 912, p. 168756, 2024.
[6] N. M. Alwi et al., "Retrofitting measures for climate resilience: Enhancing the solar performance of Malaysian school buildings with passive design concepts," in IOP Conference Series: Earth and Environmental Science, 2022, vol. 1102, no. 1: IOP Publishing, p. 012014.
[7] Antonik-Popiołek, P. (2021). Influence of solar radiation to the temperature inside a three-layer partition in winter season. Acta Scientiarum Polonorum Architectura, 20(2), 75–82. https://doi.org/10.22630/aspa.2021.20.2.16
[8] N. S. Firman, S. A. Zaki, N. W. Tuck, M. K. Singh, and H. B. Rijal, "A study on adaptive thermal comfort and ventilation in Malaysia secondary school classrooms of tropical climate," Building and Environment, p. 112701, 2025.
[9] N. Norazman, A. I. Che Ani, W. N. W. Ismail, A. H. Hussain, and K. N. Abdul Maulud, "Indoor environmental quality towards classrooms' comforts level: Case study at Malaysian secondary school building," Applied Sciences, vol. 11, no. 13, p. 5866, 2021.
[10] Y. Wang, S. Zhang, D. Wang, and Y. Liu, "Experimental study on the influence of temperature and humidity on the thermal conductivity of building insulation materials," Energy and Built Environment, vol. 4, no. 4, pp. 386-398, 2023.
[11] Hwang, R., Fang, P., & Chen, W. (2023). Impact of solar radiation on indoor thermal comfort near highly glazed façades in a hot-humid subtropical climate: An experimental evaluation. Building and Environment, 243, 110725.
[12] A. M. Al Yacouby, M. F. Khamidi, M. F. Nuruddi, S. A. Farhan, and A. E. Razali, "Study on roof tile's colors in Malaysia for development of new anti-warming roof tiles with higher Solar Reflectance Index (SRI)," in 2011 National Postgraduate Conference, 2011: IEEE, pp. 1-6.
[13] Nur'aini, R. D. (2014). Investigation of Ceramic Roof, Metallic Roof and Concrete Roof Thermal Performance in Tropical Climate. NALARs, 13(2).
[14] B. Arman Kuzubasoglu and S. Kursun Bahadir, "Flexible temperature sensors: A review," Sensors and Actuators A: Physical, vol. 315, p. 112282, 2020.
[15] Tuck, N. W., Zaki, S. A., Hagishima, A., Rijal, H. B., & Yakub, F. (2020). Affordable retrofitting methods to achieve thermal comfort for a terrace house in Malaysia with a hot–humid climate. Energy and Buildings, 223, 110072. https://doi.org/10.1016/j.enbuild.2020.110072

INDOOR AIR QUALITY LITERACY AND DETERMINANTS OF INDOOR AIR MANAGEMENT PRACTICES AMONG MYANMAR MIGRANTS IN SAMUT SAKHON PROVINCE, THAILAND

Khant Htoo[1]*, Piyapong Janmaimool[1], Orapin Laosee[1], Thunwadee Suksaroj[1], Sherry Yueh-Hsia Chiu[2]

[1]ASEAN Institute for Health Development, Mahidol University, Nakhon Pathom, Thailand

[2]College of Management, Chang Gung University, Taoyuan, Taiwan

*Corresponding email: khant.kh9@gmail.com

ABSTRACT

Indoor air pollution inside the home is a serious public health issue, especially in vulnerable communities like migrants living in crowded or poorly ventilated houses. Migrant populations often face challenges in managing indoor air quality (IAQ) due to limited resources and lack of access to reliable IAQ information. This study aimed to explore IAQ literacy and the factors that influence IAQ management practices among Myanmar migrant households in Samut Sakhon Province, Thailand.

A cross-sectional survey was conducted with 421 Myanmar migrant households between 25–30 June 2025 using a structured questionnaire. The questionnaire measured IAQ management practices (dependent variable) and potential influencing factors, including IAQ literacy, risk perception of indoor air pollutants, attitudes towards IAQ management practices, perceived self-efficacy, perceived barriers, and socioeconomic factors. Chi-square tests and multiple logistic regression analyses were used to identify significant associations.

Although overall IAQ literacy was high among participants, only specific components—such as appraisal and application of IAQ information—were significantly linked to better management practices. In the final model of multiple logistic regression analysis, perceived self-efficacy, perceived susceptibility, and household size were also significant predictors. Households with fewer members and those with lower perceived susceptibility were more likely to engage in effective IAQ practices. Other factors, including attitudes, perceived severity, all types of barriers, gender, income, education, and occupation, were not significantly associated.

The findings suggest that improving IAQ literacy (particularly appraisal and information application skills) and enhancing self-efficacy may help strengthen IAQ management practices among Myanmar migrants, even in the presence of structural and economic challenges.

KEYWORDS
Indoor air pollution, IAQ management behaviors, Environmental health literacy, Migrant household health, Housing-related exposure

Title of the Abstract: Assessment and Characterization of Airborne Microplastics Across Indoor Settings

Name: Chaithra S, Indian Institute of Technology Madras

Contact Mail: chaithra.sraju7@gmail.com

Authors: Chaithra S a, S M Shiva Nagendra a*

Abstract:

Airborne microplastics have emerged as a significant concern in recent years, particularly within the indoor environments since individuals spend approximately 90 - 95 % of their time in indoors. Microplastic presence in indoor air, not only contributes to environmental degradation but also poses potential health risks to the occupants due to prolonged exposure. This study aims to comprehensively examine the occurrence, characterization and morphological features of airborne microplastics across different indoor environments, including a 3D printing room, waste management facility, ball mill, workshop, classrooms and corridor space and an incineration plant. The highest concentration of airborne microplastics was observed in the incineration plant (9.31 MPs/m^3), followed by classrooms and corridor spaces (6.35 MPs/m3), the waste management site (5.52 MPs/m^3) and the 3D printing room (5.51 MPs/m^3). These findings highlight the variation in microplastic concentration across different indoor environments. Particles smaller than 100 µm were the most predominant across all locations, with the highest proportion observed at the incineration plant (65%), followed by waste management facility and 3D printing room (both at 42%). Particle size ranging from 200 – 500 µm were the second most abundant, particularly in the ball mill and workshop (both at 19%). Particles larger than 1000 µm size were mainly found in classrooms and corridor space (22%), followed by department workshops (20%). Across all locations, most microplastic particles were found to be in blue, black and transparent color, with the majority being fibrous in shape, followed by fragment shaped particles. Polyacrylates, polyolefins, polyvinyl-based compounds, and polyesters were the most commonly identified polymer types across the observed indoor environments. This study reveals that, the generation and emission of microplastics are strongly influenced by the primary activities happening within these spaces, which may pose a significant health risk to the occupants upon prolonged exposure.